D1265089

THE TRUTH ABOUT WUHAN

....

HOW I UNCOVERED THE BIGGEST LIE IN HISTORY

....

DR. ANDREW G. HUFF

FORMER ECOHEALTH ALLIANCE VICE PRESIDENT
AND SENIOR SCIENTIST

Skyhorse Publishing

Skyhorse Publishing books may be purchased in bulk at special discounts for sales promotion, corporate gifts, fund-raising, or educational purposes. Special editions can also be created to specifications. For details, contact the Special Sales Department, Skyhorse Publishing, 307 West 36th Street, 11th Floor, New York, NY 10018 or info@skyhorsepublishing.com.

Skyhorse® and Skyhorse Publishing® are registered trademarks of Skyhorse Publishing, Inc.®, a Delaware corporation.

Visit our website at www.skyhorsepublishing.com.

10 9 8 7 6 5 4 3 2 1

Library of Congress Cataloging-in-Publication Data is available on file.

Print ISBN: 978-1-5107-7388-2
Ebook ISBN: 978-1-5107-7389-9

Printed in the United States of America

Contents

Acknowledgments

Thank you to everyone that supported me this past year.

First, I would like to thank Scott and Marie. Without your support, I wouldn't have been able to write this book and continue the fight for the American way of life and our freedoms. Next, I would like to thank Duso for his support and goodwilled attitude, which kept me determined to fight. Thank you, Chad, for helping around our property and for garnering support from the local community in the fight against corruption and tyranny. Chris and Brock, thank you for all the technical support that you provided in fighting against the corrupt Michigan State Police and Federal Bureau of Investigations and other US government sponsored hackers. Thank you to members of DRASTIC and the scientific community that fought valiantly to uncover the truth about the origins of SARS-CoV-2; it was truly a decentralized team effort. Thank you, Caroline, Kent, and Tony for reviewing drafts of the manuscript and for pushing through repeated hacks and criminal activity by the Michigan and federal government to prevent the publication of this book. Thank you to my new friend and attorney Tom Renz—you are one the few people that I have ever met that fights for what they believe in and is tenacious enough to win these tough battles against the government to preserve our freedoms.

Behind every great man, stands a great woman. Emily, I love you.

My insurance company, United States Automobile Association (USAA), can go to hell. Your organization was a nightmare to deal with.

This book is dedicated to my son, George. I am sorry that I was not able to spend as much time with you this past year as we both wanted. I hope that you forgive me. From this point forward I am going to spend as much time with you as possible. I fought this fight against insanity and corruption because it was the right thing to do. There are some things in life that are more important that power, money, and spending time with your family. My fight against these evil people was one of them. I hope, and want to believe, that you will never have to endure the struggles and self-sacrifices that I have endured and made over the course of mine. The reality is great men are defined by what they say no to, and what they stand up against. I hope that you have courage, honor, and integrity to do the right thing, when the odds are stacked against you. I love you, be strong, be a leader.

Freedom isn't free. I paid for it.

Foreword by Jason Bashura, Biodefense and Public Health

In March 2020, my friend Andrew G. Huff texted me and wanted to talk. He was seeking a letter of recommendation (LoR) to support his readmission to the US Army to help the looming domestic "fight" on what was evolving into the COVID-19 pandemic. An excerpt to my LoR (dated March 30, 2020) states:

> When Dr. Huff's career progressed, I was amazed but not surprised by his work ethic and focus that he and his team were engaged in pertaining to Global Public Health research including, but not limited to, evaluating the impacts that worker absenteeism would have on the food supply in the midst of a pandemic.

And the letter of recommendation concluded with:

> The depth and breadth of Andrew's experience is evident in how he has applied his vast educational foundation. He has simultaneously developed expertise in supply chains, systems engineering, security, and public health, and I believe this overlap in skillset and knowledge will serve him well in this new role. I am thrilled to be able to recommend Dr. Huff for this opportunity, and I am humbled that he asked me to offer this letter of support. Knowing that Dr. Huff would be serving OUR country in this capacity to fight the COVID-19 Pandemic to me is the penultimate opportunity for him to display courage, leadership, experience, and progress in protecting the public's health and well-being.
>
> Passionate. Tenacious. Fiercely loyal. Patriotic to *our* country, in every letter of the word.

A forward, big picture thinker. A problem-solving, systems-based, public health–minded, and practically driven freethinker. Observant. Andrew Huff's character traits and skill sets are unparalleled.

From his early work with the University of Minnesota's National Center for Food Protection and Defense (NCFPD, recently renamed to the Food Protection and Defense Institute) to his deterministic risk-based work with Sandia National Laboratories, to his academic and public health veterinarian driven pursuits with Michigan State University, to the opportunities that appeared to be promising with EcoHealth Alliance, and now in private enterprising small business—I've been fortunate to have been able to learn from, appreciate, observe, and recommend Andrew's work and spirit to improve tomorrow, based on lessons learned yesterday.

Fast forward nearly two years. . . . Another text from Andrew, one that I'll *never* forget, from Friday, November 5, 2021 at 9:27 a.m.:

Good morning Jason. Have someone at the FBI that you can refer me to right now? It's an emergency.

Now, with good conscience, can any of you honestly say that's how you want a conversation to begin? Unfortunately, that's my first memory of the journey that my friend—Andrew Huff—and his family have been on for the last few months.

Donald Rumsfeld once said, and I've *often* repeated in various iterations, among other interesting lines of thinking, "… there are unknown unknowns." Well, in my mind, now, "there are things that I cannot un-know." I cannot "forget" the tribulations that Andrew and his family— and community where they live—had, have, and continue to endure since this "escapade" began.

One of my initial responses back to him—after confirming that I'd made a few phone calls and gotten him the contacts that he needed thanks to some *trusted* colleagues—was asking about his wife and young son. The goosebumps that I get—even today as I write this—thinking about how she felt, and what their little guy will *never* know about until he's able to read and understand the overt violations of not only their home and their civil liberties, but also their *world* as they know it. And why?

Because Andrew Huff knew (and knows) too much. What ensued in the following couple of days not only gave me goosebumps, but also forced me to rethink *everything* that I'd read, heard, learned, and followed since

December 2019 (when, as a public health guy, I was aware of what was going on and hoping it wasn't going to be as bad as it could be). I had worked in local, county, state, and federal public health in various anti-terrorism capacities and public health preparedness, developing a myriad of "response plans" for everything from community-based smallpox mass vaccination to pandemic influenza readiness to anthrax point of distribution (PODs) to anti-viral distribution plans based on evolving guidance from the CDC's public health emergency response teams.

How could we have "seen" this coming and not done anything about it?

Why was this research being conducted?

Who was paying for it?

Could it *really* be the biggest façade in the history of the world?

Put yourself in Andrew's position—imagine being surveilled by drones, tailed everywhere he went, and then to find out that *they* were in his house, violating his personal space to *listen* to who he was talking to, what he was saying, and when. Put yourself in his position, and open your mind to all he has to say in this book.

CHAPTER 1

Both Science & the US Government Are Broken

During the week of October 25, 2021, I decided to come forward as a material witness and whistleblower related to SARS-CoV-2.

That week my popularity on LinkedIn and Twitter seemed to be taking off, as my followings on both platforms were rapidly increasing.

Before the rapid increase, to say that I was bad at social media platforms was an understatement. Maybe it wasn't so much that I was bad at social media, but that my personality and style often come across to others as being a jerk or being mean, even though that is not my intent.

I must warn you; I will never win a Miss Congeniality contest.

My directness is due to my desire to get to the point quickly so that we can get the work done and find answers. This is likely a left-over trait that I learned in the military, where there is no time or place for beating around the bush. When there is bad news to be delivered, or if you made a serious mistake, it is best to be honest, concise, and develop a solution rapidly. Then, equally as important as developing a solution, is quickly implementing the solution. Failing fast is the most effective way to move quickly. If you made a bad decision, then you can identify the failure and correct the issue.

I am the soldier you want in your foxhole because I understand the battlefield.

Over time, and as you age, your heuristics and schemas improve from being exposed to different situations, and you become better at predicting the most likely best solution. Consequently, over time you will become better at hypothesizing the best-case solution to a problem or finding the truth. By using this process, and having a healthy dose of skepticism, I discovered, and reveal in this book, the biggest scandal in the history of the United States. I am deeply saddened and angered by the truth, and I am terrified of the direction that our great nation is heading.

This book aims to do what our leaders have failed to do: tell the truth.

During the week of October 25, 2021, I reached out to both Alex Berenson and Dr. Bret Weinstein via email and through social media platforms, and I was able to arrange telephone calls with both. Alex and I had three or four long conversations where I really dished the dirt on EcoHealth Alliance.[1]

Alex seemed to lose interest or had bigger fish to fry after I told him about the Central Intelligence Agency's (CIA) involvement with my former boss Dr. Peter Daszak, by Peter's own admission. This was unfortunate because we did not have the opportunity to get to what I felt was the bigger story related to EcoHealth Alliance. Despite the conversation ending abruptly, I really enjoyed speaking with him, as he asked thoughtful, detailed, and astute questions.

Since I can only presume that I am the first and only EcoHealth Alliance employee to come forward, Alex really was interested in the people that work there and their personalities, how I came to my conclusions and assessment of the organization (which changed and became increasingly negative over time). He is an excellent writer and journalist (difficult to find these days due to the suppression of the truth, and those that tell it, by the mainstream media). I look forward to the opportunity to hopefully meet him in person someday.

Later that week, I had two of the most stimulating scientific discussions of my life with Dr. Bret Weinstein and Dr. Jan Jekielek. Shortly after those conversations, the *Epoch Times* published an article and infographic which stated that EcoHealth Alliance had been working with the CIA.[2]

During the nearly two-hour conversation, we discussed every single aspect of COVID-19. We discussed everything from the failure of the vaccines to how the leaky mRNA vaccines were selecting for increasingly virulent and potentially greater pathogenic strains of COVID. Meaning that through a process of natural selection via reproductive fitness (Darwinian Theory of Evolution) the SARS-CoV-2 virus was becoming more transmissible and less deadly.[3]

Simply, the virus would become more transmissible because when a person is vaccinated with a leaky mRNA vaccine, the strains that match the mRNA vaccine are blocked from reproducing, and the strains in the mRNA-vaccinated person's body continue to replicate.

Thus, the leaky mRNA vaccines cause the strains circulating within the vaccinated person's body to spread to other people regardless of the other person's vaccination status (either being vaccinated or unvaccinated). This is one of many reasons why the United States' COVID-19 vaccination policy was flawed and was ultimately doomed.

The mRNA vaccine platform stands for modified ribonucleic acid. However, calling the mRNA treatment developed to treat illness due to COVID-19 a vaccine is misleading. In fact, the COVID-19 vaccine is not a vaccine at all, but rather it is a gene therapy[4]. During the time when this book was written in the spring of 2022, it was highly contentious to call the vaccine a gene therapy. Simply stating scientific facts like the COVID "vaccine" is ineffective or stating that the "vaccine" had side effects was reason enough to be banned from social media platforms like LinkedIn (a Microsoft company), Facebook, Twitter, or YouTube (a Google company). In fact, I was banned from both LinkedIn and Facebook in 2021 for somehow violating their terms of service for posting scientific facts.

I only posted facts related to science and public health, and then placed my observations about the mRNA vaccines, associated mRNA vaccine side effects, COVID vaccine effectiveness, and the origin of COVID into the appropriate scientific context.

What happened from 2019–2022 was quite astonishing: An outspoken minority of scientists was proven correct related to numerous issues related to COVID. From mRNA "vaccine" effectiveness to the laboratory origin of SARS-CoV-2, these scientists were proven correct.

Another glaring example of the minority scientists being proven correct occurred at the World Health Summit in 2021. In 2021, the president of Bayer's Pharmaceuticals Division, Stefan Oelrich, stated:

> ...Together with Bill and Melinda Gates we're working very closely on family planning initiatives .... We are really taking that leap [to drive innovation]— us as a company, Bayer—in cell and gene therapies ... ultimately the mRNA vaccines are an example for that cell and gene therapy. I always like to say: if we had surveyed two years ago in the public—"would you be willing to take a gene or cell therapy and inject it into your body?"—we probably would have had a 95 percent refusal rate.[5]

This was astonishing because it confirmed several theories that have been labeled as conspiracies. First, that Bill and Melinda Gates were heavily involved with mRNA vaccine strategy and influencing policy. Second, that the mRNA platform is, in fact, a gene therapy. Third, that the tragedy and fear surrounding the COVID-19 pandemic were used to bring the mRNA platform to market, and, if not for the fear related to COVID, the mRNA platform would have not been accepted by the public.

What was well-known by scientists and people watching the COVID story closely was now being stated publicly by one of the leading pharmaceutical company's executives.

While hearing these truths was great, they conveniently came a little too late for the people that were wrongly labeled as conspiracy theorists by US government employees like Dr. Anthony Fauci.[6][7]

The US government-sponsored Moderna and Pfizer "vaccine" trials acknowledged that their gene therapy technology had no impact on viral infection or transmission whatsoever, and that the mRNA gene therapy merely conveys to the recipient the ability to produce a spike protein endogenously by the introduction of a synthetic mRNA sequence into their body.[8][9][10][11] The cited, scientifically dense, clinical trial documents from Moderna and Pfizer were not concise or easy for many people to comprehend, so most people did not know that the mRNA jabs were not tested for vaccine effectiveness against transmission. It became painfully clear to everyone when Pfizer executive Janine Small admitted on October 11, 2022 that the COVID vaccine was not tested for transmission.[12]

Operational definitions like how we define a vaccine are incredibly important and are critical to measuring progress, defining success, and making objective comparisons in science. Operational definitions are necessary for honest debate and finding the truth.

Clearly, the mRNA platform is a gene therapy and is not a vaccine as promoted by the United States government and mainstream media, and the operational definition of what is a vaccine is being manipulated by US government officials and by the scientists that receive funding from these government agencies and officials. Obviously, there is a large incentive for people to engage in these types of behaviors.

Since 2020, there have been numerous US government-funded, peer-reviewed scientific publications that have slowly attempted to erode and change the definition of what a vaccine is.[13][14] These scientists have stated that reducing the severity of the illness is the primary goal of a vaccine, and this redefining of "vaccine" is complete nonsense.[15]

The most troubling of these recent publications are attempting to change how we frame the origin of COVID and the United States government's response to the pandemic, and were authored by scientists that have significant conflicts of interest with both the origin of COVID and the SARS-CoV-2 mRNA gene therapy platform. The scientists that have received most of the attention are Dr. Ralph Baric, who is a professor at the University of North Carolina Chapel Hill, Dr. Peter Daszak, who is the president of the profitable non-profit organization EcoHealth Alliance, and Dr. Anthony Fauci, the director of a National Institute of Health (NIH) sub-agency, the National Institute of Allergens and Infectious Diseases (NIAID).

The real question is: why are these prominent scientists attempting to redefine what constitutes a vaccine? The answer is simple: corruption.

These corrupt scientists and government officials have been getting away with these types of behaviors for years.

This is how corruption and the funding model and process in science works.

First, identify and define the problem.

Then, identify which publications and which coauthors will have the most impact on the public discourse.

After the collaborators and target publication are identified, try to arrange a phone call with one of the target publication's editors to obtain political buy-in before a manuscript is submitted.

Then, socialize the idea with your peers and select peer reviewers that you know for a fact are on your side, support your reasoning, and support your conclusions.

If possible, include coauthors from the project's sponsors to help present a "diverse" yet unified front.

Submit the manuscript to the publication and wait for reviews.

Once the manuscript is accepted, have your coauthors and subordinates repeat the thesis of the article in other forms of media and other peer-reviewed journals.

This is how scientific consensus and "fact" is established. No research is required. This is exactly how the definition of vaccines is being manipulated. This example is emblematic of how science has become corrupted with the support of US taxpayer funding. This example is representative of what is happening throughout science: corrupt scientists will do or say anything to maintain their funding. In the context of COVID, Drs. Baric, Daszak, and Fauci are attempting to avoid prison for what is discussed later in this book.

In mid-December of 2019, I became aware of the infectious disease outbreak in Wuhan, China. Interestingly, I learned of it via nontraditional means. Typically, I receive emerging infectious disease notifications from a platform called ProMED, which is operated by the International Society for Infectious Diseases (ISID). The Program for Monitoring Emerging Diseases (ProMED) is:

> [A]n internet-based reporting system dedicated to the rapid global dissemination of information on outbreaks of infectious diseases and acute exposures to toxins that affect human health, including those in animals and in plants grown for food or animal feed. Electronic communications enable ProMED to provide up-to-date and reliable news about threats to human, animal, and plant health around the world as quickly as possible.[16]

This is the tool that most epidemiologists and public health officials use to receive notification that there are anomalous health events occurring. While working at EcoHealth Alliance, ISID and ProMED were subcontractors on one of my contracts. I had the great pleasure of working with the legendary Drs. Marjorie Pollack, Larry Madoff, and the late Jack Woodall. Dr. Woodall was an early pioneer, using the internet to rapidly collect and validate health surveillance information related to emerging health threats.

Much to his credit, ProMED has been one of the most impactful health surveillance tools ever created. ProMED relies on infectious disease experts like Marjorie and Jack to identify, analyze, and request information from people physically located near the source of the event. If the people near the source of the event are compromised in some way, or are being censored by the government, then the health surveillance information (a.k.a. the intelligence) collected can be easily manipulated in a variety of ways.

When the health event occurred, where the event occurred, and the characteristics of who is affected can be presented in a manner to intentionally mislead or distort the facts. The first ProMED mail report related to COVID-19 occurred in late December of 2019. By this time, the COVID-19 pandemic had likely been ongoing for at least weeks, if not months.

Despite ProMED's past demonstrated utility and impact, it seems that the platform may not be as effective in places where speech or communications are restricted by the government. Also, the nature of how we communicate and share information has changed.

Relying on experts to analyze cases and outbreaks is no longer required for some diseases, as machine learning has been successfully used to automate the analysis of infectious disease reporting in digital and written formats.

In the intelligence community, the analysis of digital information to identify early warnings or precursors is also known as signal intelligence (SIGINT) or sometimes communication intelligence (COMINT), depending on the specific context.[17] These are some of the types of technologies or platforms that I designed, built, and refined for an alphabet soup of US government agencies over a period of six years.

***

Despite the technological advances in identifying infectious disease outbreaks, I first learned of the outbreak in Wuhan, China while cruising through a professional forum structured like Reddit.

In mid-December 2019, one of the forum members claimed that he was from the region and that there was a wide-spread infectious disease outbreak occurring in Wuhan causing thousands of deaths. The claim immediately piqued my interest, and I decided to attempt to validate the claim. First, I checked ProMED, and there were no reports or requests for information (RFI) related to an infectious disease outbreak in Wuhan, China.

Next, I thought of alternative ways to validate the outbreak information, and I recalled a method that I had learned in graduate school. During severe infectious disease outbreaks, bodies need to be rapidly disposed of for numerous health and sanitation reasons. The most common way of disposing of bodies in a dense urban area like Wuhan is cremation of the bodies. There is simply no place to bury the bodies quickly and moving the bodies out of the city creates new exposure risks and requires a new supply chain to be established. It is much easier, safer, and more efficient to burn the bodies.

When human bodies are burned at crematoriums, they release large amounts of fine particulate matter into the air, which can be easily inhaled and caught in the lungs. Particulate matter (PM) comes in all sizes but two of the most common measured by institutions concerned with air quality and pollution are PM 2.5 and PM 10.[18] The number after PM indicates the size of particle in micrometers, and high PM 2.5 and PM 10 concentrations in the air cause numerous respiratory diseases.

For these reasons, PM 2.5 and PM 10 data are collected continuously for numerous large cities or places with air quality concerns globally. Often, these air pollutant data are loaded into and modeled by a type of software called a geographic information system (GIS) so that the dispersion and concentration of the particulate matter can be visualized with a plume dispersion model. You are probably familiar with Google Maps or Google

Earth, and these are simple types of GIS platforms that are used for land navigation or simple spatial analyses, and the map layer can be overlaid with other objects or shapes to represent various phenomena like wildfire origins and smoke plumes in the western United States during fire season.[19]

Governments internationally, from small cities to large national organizations like the US and Chinese Environmental Protection Agencies, and from small private companies to large research-focused academic institutions collect PM 2.5 and PM 10 air pollution data. Many of these organizations provide the data for analysis in a GIS or provide an online GIS and plume modeling platform for PM of various sizes.[20]

Since Wuhan, China and eastern China in general have significant air pollution problems, obtaining PM data down to the city block level is not difficult, and many GIS and plume models already exist. So, in mid-December 2019, I analyzed the PM 2.5 data and created a map layer with pinpoints for the crematoriums in and around Wuhan, and I found that the crematoriums were operating in overdrive.

I was shocked, and I immediately started contacting friends and colleagues in my inner circle to share the information that I had learned. In the back of my mind, I knew that EcoHealth Alliance had been performing gain of function work at the Wuhan Institute of Virology, but that fact alone was not enough to draw any conclusions as to the source or cause of the disease at the time.

The thing that was most puzzling to me was why wasn't the US government sounding the alarm to the public and taking actions to prepare and respond to this terrifying emerging infectious disease threat?!

There have been times when governments have not been transparent about disasters with significant global health–related consequences. Classic examples are when the USSR attempted to hide a biological laboratory leak in 1977 that resulted in the re-emergence of the H1N1 pandemic flu strain which killed 700,000 people globally[21] and the explosion and meltdown of the Number 4 nuclear reactor in Chernobyl, Ukraine that resulted in radioactive fallout being spread across Europe and western Asia, which caused an estimated 200,000 to 985,000 latent deaths due to various types of radiation exposures to the population and environment.[22]

It turns out that SARS-CoV-2, the agent that causes COVID-19, will be added to that list, as will China and my home country the United States of America.

CHAPTER 2

The Long Path to Enlightenment

I am making some rather bold claims in this book, and numerous people, including Dr. Peter Daszak, president at EcoHealth Alliance, have claimed that I am "lying about everything."

I am not lying about anything.

Dr. Daszak is a liar, and I will prove it in this book. I feel that communicating my past is important because I prefer to be up-front and transparent about who and what I am, and I am not a person who glosses over my failures. Those failures are important in shaping the person that I have become, and there is no better way to learn than via failures.

In 1995, my father and I took a trip to Colorado where we made it a point to visit the Air Force Academy in Colorado Springs. At the time, I was 5'8" and had my eyes set on becoming a fighter pilot. In 1999, I grew six inches in a year to just over 6'4", and my dreams of being a fighter pilot were dashed. I didn't complete high school for a myriad of reasons. I had made the decision to complete all my core courses by the end of tenth grade, I had passed all the necessary high school graduation tests and had performed well on the ACT exams.

At the end of tenth grade, all my core competency courses in history, science, math, and English had been completed. The only courses that I had left to complete were elective courses, and I was completely bored with high school. I enrolled and participated in just enough high school coursework to stay active in athletic programs and the social life at school. At the age of sixteen, I had my sights set on college and wanted to leave the K–12 public education system as it spent most of its resources directed at the lowest common denominator of students. Later in life, I learned that a non-trivial number of people that obtain PhDs had similar feelings and conflicts with the public education system.

I wanted to be enrolled in a program called the Post-Secondary Education Option, but my administrator did not believe that I would make it through college and would not approve of me attending college, even though I'd already been accepted to several programs. So, I spent most of my time skipping classes, reading in the library, and focusing on part-time work and athletics.

In the year 2000, I decided to attend St. Cloud State University (SCSU) in St. Cloud, Minnesota. The decision was driven by the fact that some of my best friends at the time were attending SCSU and the tuition and cost of living were quite reasonable. Tuition my freshman year was only $2,784 per year, and my half of the rent was only $200 per month. At the time, St. Cloud State had a reputation as a party school.

Back then, I wasn't seeking the party atmosphere, but I was not attempting to avoid it either. It seemed at this party school that every night, except Tuesdays and Wednesdays, there was an excuse to drink heavily. Most freshmen drank and flunked their way out of college, never to be seen again on campus. I later learned in life that this is true at most universities regardless of their perceived status as a party school.

I had decided to major in finance and economics, and like most freshmen at SCSU, I developed what I eventually determined was a drinking problem during my freshmen year. I drank essentially every Thursday, Friday, Saturday, and Sunday. I also smoked two packs of cigarettes per week on average, which were quite affordable at only three dollars per pack in the early 2000s in Minnesota.

As I observed my friends at school begin to flunk out my freshmen year, I decided to take things more seriously. I was spending way too much time and effort on chasing the opposite sex and drinking.

***

In early September 2001, my roommates and I were walking on a beautiful sunny fall day to class across campus. We were joking around and making small talk, when a hysterical woman our age ran up behind us and exclaimed that an airplane had just hit the World Trade Center around 8 a.m. Central time. My roommates and I didn't think much of it and, in my mind, I instantly thought of the building that was struck with a Cessna flown by a deranged man that was upset with the Internal Revenue Service over his taxes.[23]

As the young woman ran past us, we snickered about how crazy she was. As we approached the center of campus, it was eerily quiet, and people were jogging into buildings and were huddling around the latest 1990s bulky and heavy flat screen television technology. As my roommates and I split up to go to different buildings, I had about forty minutes to burn before my first class of the day and decided to walk into the Atwood Memorial Center.

I approached one of the televisions and the building was packed full of people huddled around the brand-new HD televisions. The student hall was so quiet that you could hear the light static from a person walking across the carpet forty feet away. As I walked up to the television and began to watch the smoldering World Trade Center tower, a second passenger jet came into the frame and struck the other tower, bursting into flames.

As the tower was struck, two young women took off crying and ran out of the student center. I began to call and text friends, and the cell phone providers' networks were overloaded and many of the attempted calls and text messages failed to connect.

I couldn't believe what I was seeing and hearing and decided to go home as classes were beginning to be canceled.

I eventually connected with my roommates over the next thirty minutes, and we walked backed to our car. During the brief walk, we discussed what was now obvious: the United States was under attack, and we were entering a new period of war.

The previous year, I had met a close friend at one of the designated party houses near campus jokingly named The Ritz. The Ritz reeked of booze and cleaning solvents and was the kind of place where shoes were required to protect yourself from the filth. This new friend was wild, daring, obnoxious, boisterous, and brilliant. For this book, I will call him "Harry."

Harry was an infantryman in the Minnesota Army National Guard, which had a rich military history. The Minnesota National Guard participated in several battles throughout the Civil War.[24] In 1861, they were heavily engaged at the First Battle of Bull Run and the Battle of Ball's Bluff. In May 1862, the Minnesota National Guard became part of the First Brigade, Second Division, Second Corps of the Army of the Potomac.

As a part of this Corps, the Minnesota National Guard participated in the Peninsula Campaign, the Seven Days Battle, and Antietam in Maryland where they sustained heavy losses. These battles paled in comparison to the fighting which occurred at Gettysburg, where the First Minnesota was

crucial to the future success of the Union Army. During this second day
of fighting at Gettysburg, troops of the Minnesota Army National Guard
charged the Confederates, securing the Federals' position on Cemetery
Ridge, which was essential to winning the battle.

During the Battle of Gettysburg, the Minnesota Infantry secured
Virginia's Confederate battle flag, and the flag is on display at the Minnesota
State Capital Building. The Minnesota National Guard was called to action
for both World War I and World War II, where units served globally in most
theaters and campaigns, but would not see action again until activated to
serve in Operation Joint Forge in Bosnia and Herzegovina and Operation
Joint Guardian in Kosovo.

Immediately after September 11, Harry and I spent increasingly more
time together, and I often asked him questions about serving in the military. I also enrolled in the Army Reserve Officer Training Corp (ROTC)
program at SCSU and began physical training and coursework to become
an officer. Many of my new friends in Army ROTC were also serving in
the Army Reserves or the Army National Guard, which was highly recommended by the ROTC cadre.

While the news cycle was dominated by Al Qaeda terrorists and the
counterattack of the terrorists' home base in Afghanistan authorized by
President George W. Bush's signing of the use of force, I was debating
whether I should join the fight. In ROTC, I was exposed to people working
in almost every branch of the military, and I decided that if I enlisted in the
army, that I wanted to be with my friend Harry and do what I thought was
the most brave and difficult job possible, to be an infantryman.

In early 2002, I was introduced to Harry's Army National Guard
recruiter, and I decided to enlist as an infantryman.

In summer 2002, I had filled out all the paperwork and signed the
contracts to enlist in the Minnesota Army National Guard as an 11C (pronounced "eleven Charlie"), an indirect fire infantryman.[25] These recruits go
through the same infantry basic training school in Fort Benning, Georgia
but are split apart from the rest of the company a few weeks from the end
of infantry training to receive advanced training in the mortar weapons
platforms.

In August 2002, I completed my medical evaluation and the Armed
Services Vocational Aptitude Battery (ASVAB), where I tested very well and
had very high general technical (GT) scores. GT scores are used by the military to determine if an individual has the aptitude for various occupations
in the military.

Based on my high scores, the army officer in charge at the military entrance and processing station at Fort Snelling pulled me aside into his office and tried his best to encourage me to switch occupations into aviation or medical, which I politely declined. I wanted the tough job and wanted to serve with my friends. Later that day I was sworn into the military as an infantryman.

Upon being sworn in I was ordered to report to my readiness NCO (non-commissioned officer) at the Headquarters Company Detachment of the 1-194th Armor Battalion as soon as reasonably possible. What makes the Amy National Guard different is that you can begin training with your unit before you complete basic training. This can be a huge advantage to National Guard soldiers entering the military.

After reporting to the readiness NCO, I was provided with the annual drill schedule, selected basic training dates for May 2003, and was issued all my equipment and uniforms at my armory located in St. Cloud, Minnesota.

At ROTC, I had already learned the basics of being a soldier: Army traditions, values, land navigation, movement, marching, small unit tactics, basic rifle skills and marksmanship, command structure, and operational planning. In my guard unit, I was being taught practical skills and was assigned mostly low-level tasks as a private. Guarding ammunition and weapons systems, preparing ammunition, radio communications, and a healthy amount of cleaning and maintenance of vehicles, weapons, and facilities.

During my first full drill in September 2002, I spent the day with both the scout platoon and the mortar platoon preparing ammunition for crew-served weapons training and testing. The first two days were quite boring, but on the last day both platoons had completed training and testing on several machine gun platforms: the M60 (7.62 mm), 240B (7.62mm), M2 (.50 caliber), and the SAW (5.56mm). The NCO in charge at the firing range communicated that we had two extra pallets of ammunition that had to be expended and that the M60s were to be destroyed; decommissioned and replaced by the 240B.

At the end of National Guard drill weekend, everyone is typically exhausted as the soldiers must complete a month's worth of active-duty training in a period of only two to four days. Typically, the command gets little sleep discussing the operation and completing administrative duties, and the soldiers have duties non-stop from 4 a.m. to 11 p.m., if everything goes according to plan. This meant that the higher-ranking soldiers were not in any mood for more training.

Since I was the only green soldier who had not been to basic training, I was "asked" if I wanted to shoot and receive training on these weapons systems from the most well-seasoned and disciplined NCOs in my unit. This became one of the most valuable and exciting training days of my life. I had the opportunity to receive one-on-one training, and fire through tens of thousands of rounds on a pop-up range.

I was quickly taught how to clear jams and double feeds on the crew-served weapons. Especially on the M60s as the frequency with which these machine guns jammed was part of the reason that they were being phased out of the military. I learned to fire a six to eight rough burst of fire precisely and reliably into the target, learned how to walk rounds below and into the target, and perform volley fires with a team of machine gunners to maintain a constant rate of suppressing fire during reloads, jams, or tactical movements.

At the end of our training, we fired so many rounds through the M60s in succession that the barrels became so hot that they glowed, became translucent, and began to sag, effectively damaging the M60s beyond repair.

Between August and May 2003, I continued with my course work at SCSU, and my grades were improving across the board.

By the time I attended basic training in the summer of 2003, I had completed most of the training that I would later be taught and had received advanced training in numerous areas like radio communications, night vision and vehicle operations, crew-served weapons like the Mark-19 grenade launcher, and mounted land navigation.

I began collecting every army field or technical manual that I could get my hands on and began reading them all. I was reading instruction manuals on demolition, leadership, tactics, and even field sanitation. In May 2003, I moved all my personal items into temporary storage and shipped off to basic training.

The National Guard is different than active duty in one substantial way: I trained on all the individual and crew-served weapons before attending basic training. This is a huge advantage and makes basic training very boring.

Basic training was exactly what I expected. The only personal problem that I had with infantry school was that I was bored. The classes were being taught at a very basic level, and I understood why the drill instructors needed to spend so much time instructing the recruits on basic tasks. The tasks were dangerous, and small mistakes by anyone in the unit could get everyone killed, even in training.

There were a few other National Guardsmen in my company that were also in ROTC, and we all commiserated together. Numerous people washed out from infantry school, and by my estimation, 20 percent of the recruits were eliminated due to behavior problems, mental health issues, physical injuries, or performance issues. I sustained stress fractures in my feet, a common problem at infantry basic training, which luckily healed a few weeks before graduation. By the end of training, it was impressive to see how much we all improved as individuals and as a team.

By the end of infantry school, you realize that the entire process is a mental game through which you must persevere and excel. The thing that I learned is how unprepared mentally I was before my enlistment. Near the end of training, I realized that the drill instructors had one of the most difficult missions in the military. They had to take hundreds of young boys who were not raised properly by their parents or by society and teach them how to be men.

I was no exception to this and had great admiration for my instructors for what they had to endure from the recruits, and for the wisdom they instilled in each one of us.

CHAPTER 3

The Hon'

Near the end of infantry school, our drill instructors changed their behavior and attitude toward the recruits as they now saw us as infantryman.

This included more discussions in which we engaged in dialogue with our instructors, most of whom had been combat deployed as far back as many of the 1980s US military skirmishes in Central America.[26] The instructors directly communicated that we would likely be deployed to the Middle East, as the military engagement in Iraq, of all places, seemed to be escalating by the day. Rivaling militias were fighting for power and control of resources along sectarian and cultural lines.

I felt that the instructors almost viewed us as their children, trying to provide us with the best information and any knowledge that they could that would increase our chances of survival. My company, Bravo 2-19th Infantry Regiment, completed infantry basic training in August 2003.[27] Upon graduating, I returned home to Minnesota and promptly re-enrolled in classes at SCSU.

I quickly found an apartment to rent near campus and moved my belongings out of storage and into the apartment with my friend and brother-in-arms Harry. Harry was proud of my graduation from infantry school, and we often discussed military life, tactics, and soldiering. We were both highly competitive and often practiced hand-to-hand combat with each other.

My fighting skills had drastically improved as our bouts often ended in stalemates without the use of makeshift weapons. I attempted to convince Harry to join ROTC, but he did not want to become a political manager and only wanted to be a soldier. There are many people who hold this belief and I understand why.

Being an officer involves much mental planning, writing, and sitting in briefings. As an enlisted soldier, most of the work is focused on training, execution, and the health and welfare of the unit. Officers are also held to a higher behavioral standard, and once you become one, there is less tolerance for wild escapades and hijinks—behavior that is commonplace and is almost a rite of passage among young infantrymen.

Shortly after our September drill weekend, which was typically a live fire drill weekend for the scouts and mortar platoons, I received a phone call from our readiness NCO. He initiated the code conversation indicating that I was being activated for a deployment. I received the call at about 8 p.m. on a weekday night, and Harry was standing in front of me when I answered the phone.

Embarrassingly, I couldn't remember the challenge password response to say on the telephone, since my heart was pounding, and I had butterflies in my stomach. Despite not recalling the correct challenge phrase, the readiness NCO stated, "According to US Code Title 10 you are hereby being activated for active duty as part of Operation Enduring Freedom (OEF)," and I was ordered to immediately report for duty with Harry a few days later.

The readiness NCO paused and asked me, "Do you have any questions?"

I paused, and then I asked, "Where are we going?"

In my mind, there were only two options, Iraq or Afghanistan, and since he said OEF that likely meant Afghanistan.

The readiness NCO began to laugh and stated, "Honduras."

I replied in shock and confusion, "Honduras? What the hell is in Honduras?"

Harry looked at me, shocked and puzzled.

The readiness NCO chuckled and said, "See you guys in a couple days. It's a good deployment."

After I hung up the phone, Harry's phone rang, and it was the readiness NCO. He still had to contact Harry directly as a formality. It was a quick call and no further information about our activation and deployment was provided.

That weekend we were re-assigned from the Headquarters Company Detachment (HHC-Det) to C-Co 1-194 Armor Battalion. A platoon-sized element was formed for the deployment consisting of about forty men with various training and skillsets.

We were mainly assigned communications, supply, armor (M1 Abrams tank crewmen), cavalry scouts, and infantrymen. Our newly assigned first sergeant (E8) came from outside the command and was a military police officer and criminal investigator. That weekend we began our pre-deployment checklist to ensure that we were eligible to deploy.

Unfortunately, Harry had recently had a minor behavioral infraction as a civilian, and he was deemed undeployable. I was told in the same conversation that the reason I was being sent was that I was the highest-achieving, lowest-ranking man in the battalion.

Thanks, I guess?

Both Harry and I were upset that we would not be deploying together. That weekend we were issued official US government passports, which granted us diplomatic immunity, and were also briefed on the lack of a Status of Forces Agreement (SOFA) in the country that we were deploying to, Honduras.

A SOFA is an agreement between the host country and the foreign nation stationing military force in that country.[28] SOFAs are often included, along with other types of military agreements, as part of a comprehensive security arrangement. A SOFA does not constitute a security arrangement; it establishes the rights and privileges of foreign personnel present in a host country in support of the larger security arrangement.

Under international law, a status of forces agreement differs from military occupation. A SOFA is intended to clarify the terms under which the foreign military is allowed to operate. Typically, purely military operational issues such as the locations of bases and access to facilities are covered by separate agreements. A SOFA is more concerned with the legal issues associated with military individuals and property. This may include issues such as entry and exit into the country, tax liabilities, postal services, or employment terms for host-country nationals, but the most contentious issues are civil and criminal jurisdiction over bases and personnel.

For civil matters, SOFAs provide for how civil damages caused by the forces will be determined and paid. Criminal issues vary, but the typical provision in US SOFAs is that US courts will have jurisdiction over crimes committed either by a servicemember against another servicemember or by a servicemember as part of his or her military duty, but the host nation retains jurisdiction over other crimes. In context, this meant that if we were detained by the authorities in Honduras, we would be subject to their legal process and that the US government had no right to intervene.

More simply, if there was an altercation or any legal issue involving US service members, the US government would attempt to extradite us out of the country as fast as possible. I found this to be strange, but I was young, new, and learning, so I kept an open mind.

Despite learning about our precarious position while deployed to Honduras, the command was tight lipped about what we would be doing in Honduras. We received our written orders, and I began the laborious process of withdrawing from courses at SCSU, made plans with my landlord to store my property, and set up automatic payment methods for my bills.

We were first sent to Ft. McCoy, which is in the central part of Wisconsin. Ft. McCoy was a dump of an army facility. Not that many of the other army facilities that I visited before were much better. We were assigned to living quarters that looked like they were built during World War II, which had very little insulation and winter was rapidly setting in. If there is one thing enlisted soldiers do, it is bitch and complain about whatever the current failings of the army are. We were provided with a two-month training schedule for unknown activities in Central America.

Of note, I was assigned to Combat Lifesaver Training where I was taught numerous advanced medical skills, including administering IVs, inserting breathing tubes, and treating other complex wounds or injuries, like sucking chest wounds.[29] I really enjoyed the additional medical training and took the added responsibility seriously.

Every day someone asked about our mission, and the command would not provide us any information until everyone in the unit received their interim secret security clearances.

Most of the other units at Ft. McCoy were deploying to either Iraq or Afghanistan, and the training cadre didn't really know what to do with a unit deploying to Central America that couldn't discuss what they would be doing. We went through a mix of the training lanes related to counter insurgency tactics, urban warfare, improvised explosive device response, and suicide bomber interdiction and response among the standard hand-to-hand combat training with knives, rifles, and pistols.

Additionally, our first sergeant taught us about law enforcement skills, military law enforcement and process, and criminal investigation, all of which would come in handy later in my life. We learned how to preserve, document, and collect evidence for criminal investigations and interview suspects to obtain information and intelligence which could be used for a wide variety of purposes.

After two months of training on a wide variety of skills, many of which were not specific to combat and were more akin to working with law enforcement officials, we finally started to receive tidbits of information related to our mission in Honduras. We jokingly referred to Honduras as "the Hon," mimicking the veterans that served in Vietnam. Although, serving in the Hon was a vacation in comparison to Vietnam.

From Ft. McCoy we traveled on an Air Force C-17 Globemaster, a large, four-engine jet aircraft, to Charleston Air Force base where we were to await further orders and to arrange transportation to Soto Cano Airbase, which was also known as Joint Task Force-Bravo (JTF-B).[30] Soto Cano Airbase was the largest runway used by the United States to launch missions deep into South America, as well as throughout Central America and the Caribbean Sea and significant strategic importance for this reason.

To add to the confusion, JTF-B was also known by another name, Palmerola Air Base. Whether it's called JTF-B, Soto Cano Airbase, or Palmerola Air Base (one of the few military installations where the US flag does not fly and is controlled technically by the Honduran Air Force), it has been a launching point for numerous clandestine missions throughout its history.

The installation has played a critical role for Central Intelligence Agency (CIA) operations and US military operations throughout the region since 1981.[31] Palmerola Air Base was used to deliver medical and military aid to the Contras as part of the Iran-Contra Scandal facilitated by the exonerated Colonel Oliver North.

Unsurprisingly, during my time at JTF-B, the base was still being used for overt and clandestine operations.

After three or four days of waiting to catch an air force bird directly to JTF-B, the army changed its mind and decided to send us on commercial civilian aircraft to Tegucigalpa, Honduras where we would be transported by bus to JTF-B.

As I recall, we flew from the Charleston Airport to San Salvador, El Salvador and then caught a connecting flight to Tegucigalpa. The landing at the Tegucigalpa International Airport is a real nail biter. The airport is surrounded by mountains on all sides, and the airport is located on the only flat piece of earth in the valley. The aircraft must fly fifty feet over protruding rocks and shanties to land on what is a very short runway.

Upon landing, all the civilians clapped and cheered, and we gathered our belongings to clear customs, where we pulled out our new diplomatic

passports and were waved through. For anyone that has traveled internationally and had to clear customs, it was an amazing experience in comparison.

In Tegucigalpa, we were picked up by school buses with armed escorts
and began our drive to JTF-B. As we left the wealthier, central part of the
city, I was shocked by the poverty that I saw.

The smell of hot and burning garbage and strange noxious chemicals, as
well as the sight of a man defecating out in the open on a city street littered
in garbage, was no laughing matter. The poorest of the poor found ripped
plastic sheets and dropped them over logs to make makeshift tents in the
dirt. As the sun set on the drive, you could see the sides of the mountains
burning, presumably being cleared for agricultural use.

Finally, upon arrival we were assigned sleeping quarters in something called a hooch, a wood, kerosene-soaked building on stilts with a
metal roof. They were soaked with kerosene to prevent termites and other
insects from destroying them and were on stilts due to the dangerous flash
flooding that could occur during the rainy season.

The next day we were finally fully briefed on our mission at JTF-B. Our
primary mission was to provide security for the base and to serve as the
base's law enforcement. The physical perimeter of the base was frequently
breached by the locals who would do things like steal the lights off the
perimeter fence or bicycles from US service members.

The more daring thieves would often attempt to enter buildings and
ignore the warning signs to not enter something called a VORTAC, which
is a radio frequency device system to help pilots navigate their aircraft. This,
of course, came at great risk to the health of the criminals that entered the
VORTAC while it was in operation, due to the immense amount of radiation that it emits. The security shifts were mostly boring, and most of the
action occurred at the base's entry control point (ECP) post's deputy commander for failure to abide by the post commander's off post policy.

Often, the Hondurans working on the base would attempt to smuggle
goods like consumer electronics off the base since the goods purchased at
the post exchange were much more affordable than the local markets' prices
due to tariffs and shipping costs in the local economy. Occasionally, we
would find a weapon being smuggled onto the base, or merely forgotten
in their vehicle during searches. When guarding the flight line, you would
really begin to understand the true purpose of the base.

My first shift guarding the flight line, a strangely marked, heavily
modified C-130 landed and taxied to its final stopping point. Twelve or so
burly men and a few that looked like normal US government employees

stepped out of the aircraft. Inside the aircraft were sophisticated weapon systems and electronic equipment, and I was told that there were explosives aboard. This indicated that we had to establish an extra perimeter of security for the aircraft.

I wondered who these people were and what they were doing at our airbase. The next few days, more of the potential missions were discussed. We learned that we would be periodically attached to other missions in the region and that we were the quick reaction force for the area of operation.

Quick reaction forces (QRFs) are the response teams that are called when the US government needs emergency help. They are like a 911 emergency response for the military or other US government assets. The only time that a QRF was activated during my time at JTF-B was to protect US interests in Haiti during the 2004 coup d'état to remove President Jean-Bertrand Aristide.[32]

In addition to our QRF role, we were attached to a unit commanded by an Army Special Forces captain, where it seemed that he had carte blanche to assemble teams from any of the assets available at JTF-B. Some of these missions were providing humanitarian assistance and medical treatment to remote and poor communities. Sometimes these missions were in coordination with the United States Agency for International Development (USAID). According to USAID:

> USAID is the world's premier international development agency and a catalytic actor driving development results. USAID's work advances U.S. national security and economic prosperity, demonstrates American generosity, and promotes a path to recipient self-reliance and resilience.[33]

Sometimes, it was our mission to provide training to local law enforcement or to foreign militaries. Often, they were conducting narcotics interdiction missions throughout Central America and many people from my platoon wanted to be assigned to these missions.

Rarely, we would fly in Blackhawks or Chinooks and drop off plain-clothed US government employees at strange places. I strongly suspected that these people were CIA operatives, although I never asked. It was obvious in my opinion.

We came to realize that we were the muscle for whenever our country required it, and we had no problem with that. We provided personal security details to American ambassadors, high ranking military members and their families, or other personnel that required it when asked by the command.

On one occasion, an armed robber with a fully automatic assault rifle stopped a US-owned bus and robbed everyone on the bus at gun point. We worked with the corrupt local law enforcement in Comayagua to identify the guilty party, to no avail, and then started riding the bus, armed and in plain clothes, to capture or kill the assailants, but unfortunately the opportunity never presented itself.

This was one of many incredible learning experiences about how other parts of the world functioned with government corruption daily.

Only a month into our tour of duty in Honduras, I had my first eye-opening experience about the reality of geopolitics in Central and South America. One night, after my security patrol shift ended, I decided to go off post to a bar in the nearest town Comayagua. Comayagua was representative of many tier-two cities in Honduras.

There were several large multi-national corporations that were manufacturing things like chemicals for soap or were in the fertilizer manufacturing and agricultural businesses in the Comayagua Valley.

The valley itself was at a high elevation for Honduras and had a very distinct hot and dry season followed by a tropical wet season beginning in late May. Many of the residents of the valley worked in fruit and vegetable production with three highly productive growing seasons that spanned the entire year, worked in manufacturing jobs at the factories, or, if lucky enough, worked on our military installation.

Honduras was a very dangerous place and so was Comayagua.

People were shot or killed on a frequent basis, and it had one of the highest murder, and violent assault rates in the world. The gringos (US Army and other foreign military uniforms were green camouflage and they wanted us to go home, hence the term's origin), as they referred to us, were typically viewed as off limits by the criminals, cartel members, gang members, and the police as they did not want the US government getting involved in the potential fallout from an assault or other harmful act against a US service member.

Also, we stuck out like sore thumbs. Nothing screams gringo more than a bunch of very tall, blue-eyed, blonde men with crew cuts and Midwestern accents going for a stroll in the worst parts of Central America. There was no blending in.

Once a month, our team went with other leaders from the base, including the J2 (military intelligence of a joint operation), and would visit, inspect, and analyze the neighborhoods and establishments in Comayagua and other cities, sometimes in partnership with the State Department to create risk assessments for US government personnel operating in the country.

Often, the high risk or safe areas on the map were arbitrary due to the highly variable security conditions and the requirements to make the risk maps easy to understand by personnel only glancing at them for a few seconds. That night, I decided to go to a bar that was deemed to be "safe" alone. I received my off-post pass from the security desk sergeant and proceeded to hop in one of the cabs that were always waiting at the front gate.

The twenty-five-minute drive was always terrifying and exciting on the lawless Honduran highways. The cars were in such disrepair that dangerous mechanical failures at speed were common.

Halfway through my second *ponche*, a fruit-and-rum-based cocktail, a scruffy, overweight white man in his late forties wearing a baseball cap walked into the bar, sat down next to me, ordered a beer, and began to make small talk in perfect English, so I presumed that he was an American.

After fifteen to twenty minutes, the man started to try to recruit me for a private security operation in Africa, which I politely entertained while privately thinking the man was crazy.

He was offering me $150,000 to guard diamond mines in the Congo. Next, as we continued to drink and chat, he started asking questions that, if answered, could have been used to ascertain my base's force strength, capabilities, and missions.

This sudden change in conversation set off all the red flags from the counterintelligence training I had received. I quickly made up an excuse as to why I had to leave, hopped in a cab, and returned to base.

Upon returning to base at roughly 11 p.m., I immediately reported the incident to my squad leader and was told to report the incident to the base J2. The security desk sergeant called the J2 in his hooch, roused him out of bed, and then we briefly met and spoke with each other, where I gave him a full report of what had just happened.

He thanked me, told my squad leader that I had done the right thing, and told me to report to the Secured Compartmentalized Information Facility[34] (SCIF) to be debriefed at 9 a.m. I went back to my hooch, went to sleep, woke up early for physical training, and then put a uniform on, even though it was my day off, to report to the SCIF.

Upon reporting to the SCIF, I was escorted into a conference room next to the J2's office where the army lieutenant (J2) debriefed me on what had happened. He told me that the man was a well-known Chinese spy.

I was shocked.

Apparently, this Chinese spy had been operating in the area for over a year. The lieutenant then explained to me that Central America and Honduras is a hot bed for foreign spy activity because of all the foreign governments fighting for influence and resources throughout Central America.[35]

After his briefing, I then answered specific questions about what had occurred while the J2 and a man dressed in civilian attire, that I did not recognize, took notes. After the briefing and back-briefing concluded, I was released for the day.

After these events, I never looked at Central America or global foreign powers the same way.

Every time I noticed a bridge or school being built in Honduras, I wondered who was paying for it and what their true objective was. This is the reality in third world countries with vast natural and human resources.

These real-world experiences, combined with the continuous military training, would significantly aid my survival in the future.

* * *

During that same period, we were introduced to Special Forces Captain "Wally." Captain Wally would be our leader on numerous missions and was excited that we wanted to learn his advanced unconventional methods. He immediately seemed to be fond of my squad, and we began training with him daily in advanced hand-to-hand combat techniques.

We were like sponges and eagerly awaited his personal training on our days off from security related duties. The man was a machine and seemed to know every tactic or trick you could imagine. During one of these training sessions, one of my squad members snapped his tibia in grappling training during a flip and leg bar maneuver, and I had to run frantically two miles to obtain emergency medical assistance.

The medics raced to treat the man, and he was almost instantly medically evacuated by helicopter back to the United States. We didn't stop training but received a stern talking to from our actual infantry company commander about training too hard, which I later determined was representative of the typical and constant irony a person would experience while serving in the US military.

A few days later, Captain Wally stopped by my hooch with my squad leader to inform me that I had been selected to go on the next counter narcotics mission. I received a crash course in narcotics smuggling and a classified briefing about the mission (the mission has since been declassified).

The process is rather simple: cocaine or other contraband travels north predominantly from Colombia and cash and weapons flowed south. At the time, the FARC[36] (a.k.a. Fuerzas Armadas Revolucionarias de Colombia) and northern cartels of Colombia were prominent players in drug trafficking. Drugs are transported three ways—by trucks, by air, or by sea. At the time, narco-submarines did not exist so go-fasts were typically high-powered racing boats or were civilian aircraft that would take off from Colombia and fly north while zigzagging (thinking that they could avoid the United States' sophisticated radar systems). In fact, these flight maneuvers often confirmed our drug smuggling suspicions.

There is only one road that easily connects Central America to the United States and that is CA-5. CA-5 is the road that JTF-B is located on, so hunting for drug smugglers merely required setting up checkpoints on the highway and using canines to identify the drug shipments. For anyone that has driven through Central America, moving large quantities by truck would be a long, slow, and expensive proposition for the drug smugglers, as there would be numerous military and law enforcement officials that would have to be bribed.

Therefore, most of the drugs are sent via airplane or by boat.

The first narcotics interdiction mission was to fly from JTF-B to a military installation in Guatemala to intercept airplanes loaded full of cocaine that were landing on the border with Mexico. Since Mexico has a real air force, comparatively speaking to their Central American counterparts, they will shoot down smugglers if they violate Mexico's airspace.

Being keenly aware of this, the drug smugglers will land or crash their planes in swamps on the Guatemalan side right next to the Mexican border and then hundreds of people will run out to the airplane and grab whatever cocaine they can and then run into Mexico where the cocaine is then repackaged to be smuggled into the United States.

Before departure, members of the Drug Enforcement Agency (DEA) and Federal Bureau of Investigation (FBI) joined us with their spiffy brand-new M-4 rifles and camouflage uniforms that had never touched dirt. The soldiers and airmen were carrying GAUs, fully automatic short-barreled rifles, and pistols. One other plain-clothed individual caught a ride with us and immediately departed when we landed near Flores, Guatemala.

The Guatemalan special forces provided us 24/7 security at our hotel, which was in a touristy area of the city. It was hardly clandestine. The next day we found an area that was like where the mission would be the next day. We rehearsed the assault from the Blackhawks no less than seven times.

As we approached the landing area, we were to jump out of the helicopter, secure the area and aircraft containing the cocaine, and then assault the opposition in four-man fire teams. I wasn't scared of the drug smugglers. I was worried that the DEA or FBI agents tagging along would be a liability or would accidentally shoot one of us.

That night the low-ranking soldiers and airmen assigned to the mission, including myself, had to guard the helicopters and man the tactical operations center (TOC) while the others went to the Guatemalan equivalent of a gentlemen's club. Late at night, while we were manning the TOC, the door flung open and startled us so badly that we almost shot a member of the Guatemalan Air Force.

He came in the door shouting "Track!" and pointing to the sky.

It took us a second of staring at each other, trying to decipher what he was saying, and then it clicked. He was trying to tell us that they were tracking a cocaine-smuggling aircraft coming into our area of operation.

We ran back and looked at our computer that was connected to the Airforce AWACS, an airborne radar system, and we saw the same thing that they did. I immediately called Captain Wally on the satellite phone to give him a situation report, and my other teammates began to prep the weapons and the helicopters.

After updating Captain Wally, he said, "We're going to let that one go by. Stand down."

I was so shocked that I asked him to repeat the order, and he gruffly said, "Stand down!" and hung up the phone.

I was dumbfounded, and I repeated the order back to everyone in the room who then looked as confused as I felt. I simply said "No" to the Guatemalan airman, and he replied "No?" with a quizzical look on his face, and I repeated the statement, "No." He stormed out of the TOC.

I had just received one of the best international political lessons in the world. The US government and global superpowers can influence who is in power by simply not enforcing certain laws. In essence, this allows a person, organization, or political group that the superpower "likes" to engage in "illegal" business activities. This indirectly provides financial support to the preferential person, organization, or political group, without any direct financing or support from the superpower.

The next day we received intelligence that our mission cover had been blown, and I didn't find this surprising. We were not attempting to be sneaky. In fact, we were behaving oppositely.

We flew out to the go-fast landing area on the border and flew circles over dozens of heaps of mangled aircraft that had made their final cocaine delivery.

There were several narcotics interdictions with various outcomes. I was seeing and learning first-hand the complexity of US international relations and policy and the implications of those policies.

I realized on the flight home from the mission in Guatemala that the war on drugs was not a war on actual drugs.

The war on drugs was a mechanism for the United States to extend physical force and influence international policy.

The commander telling us to stand down on that mission had nothing to do with the drugs. Standing down and allowing that airplane to land and unload its cargo was about enabling someone the United States liked, or perhaps viewed as the lesser of two evils, to do business. Stopping cartel members, or the politicians in Central America that relied on their funding, was more obvious and dangerous than simply allowing some to stay in power by allowing them to do business.

Much of the intelligence collection and operations with USAID and the CIA, as I saw it at the time, were more about building friendships, alliances, and helping our allies stay in power.

Often the CIA agents were just quiet, listening, and observing.

When they spoke, their statements were wise and precise.

Our country's efforts in Central America, based on my limited experience and exposure, seemed clean and efficient.

Leading up to my OEF deployment, President George W. Bush was touting the "Coalition of the Willing" to fight the war in Iraq. This coalition included support from forty-nine countries globally, one of which was Honduras.

We had watched four hundred Honduran Army soldiers take off JTF-B to go fight in Iraq, only to watch them return and almost crash land about a week later. The aircraft was having trouble with its landing gear and had to circle the valley at least five times before making the final approach.

Luckily, the airplane's landing gear did not fail when it touched down on the final approach. What was not reported in the mainstream news was that a credible threat from Al Qaeda had been levied against the Honduran government and elite, and the terrorists were threatening to blow up malls and movie theaters in the wealthy financial district unless the Honduran authorities withdrew their forces from Iraq.

The politicians caved to the terrorists' demands and brought their soldiers home from Iraq shortly after the demands were levied by Al Qaeda.

One night after concluding my security patrol, I sat down to dinner and CNN was televising the war in Iraq. A heavy engagement had just wrapped up in the Anbar Province near Fallujah. The CNN reporter was interviewing soldiers about the fight, and I was shocked as the soldiers that were being interviewed were soldiers from my platoon at basic training.

They did not look emotionally well, but they were alive and looked physically fit, so I guessed all was OK, relatively speaking.

In 2004, the military enacted a program called Stop-Loss, which prevented people from exiting the military and essentially allowed the government to violate their contracts with service members.[37]

This program was later ruled to be illegal, but it was very real at the time.

Also, the army had a policy that reservists and National Guard members could only serve up to twenty-four months of active duty per enlistment and could not be ordered to active duty without at least twelve months between deployments. Both policies, while well intended, were terrible for the soldiers and force retention.

Near the end of our tour, we received word that our battalion back home in Minnesota was preparing to deploy a company to Iraq. I felt that there was nothing worse than being on active duty for a year, then returning to normal life, to only go through the same process again twelve months later. I also felt terrible that I was somehow lucky enough to be on this incredible dream deployment, while my fellow soldiers were fighting in Iraq. I loved being on Active Duty in the US Army, I loved serving our country, and I felt a great sense of honor serving our country. Mentally, I decided that I wanted to serve as an active duty commissioned officer.

At the time, I was against the invasion into Iraq, but I felt a duty for the occupation and recovery of Iraq to be successful. Due to all these facts, and my sense of duty and service to the country, I requested a meeting with my company commander and first sergeant in Iraq and volunteered to deploy and fight in Iraq.

I felt it was best to get it over with and it was better to pick your poison, rather than to be assigned to a unit randomly.

CHAPTER 4

Playtime in the Sandbox Begins

I could write an entire book about my time in the Army, and I have intentionally only selected a few of the many stories from my service to demonstrate my experience and knowledge as it relates to the content of this book.

After returning from Honduras and being discharged from active duty in the summer of 2004, I felt proud, seasoned, and tough. Having recently been promoted to private first class, I was eager to earn my next promotion, and the reality is that volunteering for deployments was the fastest way to advance in rank as the rules for promotion during deployments and war were less prescribed and were more flexible.

I spoke with Captain "Anzer," who would soon be my new company commander. He informed me that his Company Alpha (A) Company 1-194th Armor battalion would be deploying to Iraq and that they were in the process of reassigning me to A company. He thanked me for volunteering to deploy to Iraq, along with two other outstanding soldier volunteers that I had just recently served with in Honduras, and communicated my next report date to the new unit, which was in September 2004.

This gave me a few months to relax, enjoy the summer, and physically train for my next deployment. When my parents heard the news that I volunteered, they were devastated, and my friends thought I was nuts.

Not many civilians could or would understand my reasoning for volunteering.

The best part about the deployment to Iraq was that many of the men from my home platoon in the headquarters detachment were being ordered to active duty, along with my friend Harry. He was able to remedy his behavioral infraction for an Iraq deployment, or perhaps the standards were reduced, due to the increasing operation tempo and demands being placed on the National Guard to sustain the Global War on Terror along with the Iraq Campaign.

Regardless, I was pleased to be deploying with one of the best infantry-men I knew.

First, we began training at Camp Ripley, MN. By any branch of the US military's standards, Camp Ripley is an incredible National Guard training facility. It has state-of-the-art training facilities and firing ranges and was well known for Winter Warfare Courses during the harsh -30° F winters. It is designated the primary winter training site in the United States. Foreign military units from Canada, Great Britain, Norway, and the Netherlands conduct training exercises on a regular basis at Camp Ripley.

After about a month of conducting training at Camp Ripley in mostly urban warfare, vehicle maneuvering, cordon, and searches, we deployed to Ft. Dix, New Jersey. Ft. Dix was an old Army installation. Our living quarters were modern, although we mostly were engaged in a month-long field training exercise at a pretend forward operating base (FOB) that was to mimic missions and conditions in Iraq.

The weather was terrible in late November and December, with cold rain or sleet for most of these months. Jokingly, in the infantry we would say, "If it ain't raining, we ain't training."

The joke was on us though.

We were issued broken and heavily abused equipment and vehicles that, by army definition, were not serviceable. We lived in old Korean War–era canvas tents that heavily leaked water and were heated with barely functional and terrible-smelling liquid fuel heaters, and were generally cold, filthy, and wet during most of the training.

The "combat" training instructors at Ft. Dix were mostly army reservists with backgrounds in non-combat specialties or who had no deployment experience. Despite these numerous shortcomings and difficulties, we made the best of the training.

Due to my past deployment experience, and my gung-ho infantryman attitude, the company's leadership seized on my high level of motivation and experience and began to have me assume roles training the other men in the unit on various tasks and skills and assigned me to leadership duties above my pay grade.

For example, on one occasion I was tasked with leading roughly forty people up to the rank of staff sergeant on a live fire training exercise on a pop-up target range for the squad automatic weapon (SAW), which is a machine gun that is light enough to be carried and shoulder fired at a rate of eight hundred rounds per minute. It is a highly effective and reliable light infantry weapon and was my personal weapon of choice on the battlefield. I lovingly referred to mine as "The Pig," which I customized with a collapsible stock, short barrel, forward pistol grip, and had the ability to quick connect laser pointing sights, infrared, or night-vision sights.

One of the first difficulties I experienced in this role is that the range cadre at Ft. Dix, a master sergeant (E8), refused to accept that I was in charge at the rank of specialist and attempted to undermine my position and authority that had been granted by my command. So, this person decided to try and make my life hell by constantly providing different directions to my "subordinates" and me.

Nevertheless, I dealt with this asshole, and was able to accomplish the mission of getting the forty soldiers qualified on the weapons platform. I didn't ask for the extra responsibility—I was ordered to do it—and it is much more difficult to effectively lead in the army if you don't have the corresponding rank.

Due to my performance and leadership at Camp Ripley and Ft. Dix, I earned my first medal—the Army Achievement Medal.

We were all eager to leave Ft. Dix and deploy to Iraq. We learned a few weeks before Christmas that we would be flying to Ramstein Air Force Base, then to Camp Victory in Kuwait, and we would be briefed on our duty location and mission in Kuwait.

During the last few weeks before deploying, many of the men were living like it were going to be their last days on Earth.

Some drove and crashed rental cars like they were race cars, and some others maxed out their credit cards or took out loans and spent the money buying gifts for their girlfriends, patronizing prostitutes, consuming alcohol, or gambling in Atlantic City.

It is the kind of behavior that is quite predictable for combat arms soldiers deploying to a war.

In late December, our company boarded buses and drove over to the other half of the military installation, McGuire Airforce Base, where we promptly loaded and boarded a chartered civilian jet to Dublin, Ireland. It was strange to be walking onto the aircraft with rifles, pistols, and machine guns, and we all made coy jokes about it with the civilian flight crew.

We came to learn that they were flying people back and forth from Iraq and Afghanistan daily and had been doing so for about the past six months, as Operation Iraqi Freedom was approaching its third year and Operation Enduring Freedom was approaching its fourth year.

As the aircraft approached Dublin, I had never seen greener grass in my life in the middle of winter. We had about an eight-hour layover where we hopped on another chartered flight to Germany, and then after a long layover in Germany, we boarded another chartered flight to Kuwait. In total, the travel time from the United States to Kuwait was a miserable twenty-two hours.

Kuwait was surprisingly humid, scorching hot during the day, and cold at night. The country was covered in fine, powdery sand that found its way into every nook and cranny that you could imagine. In Kuwait, we were to complete more live fire training and receive our final Iraq-specific briefings before deploying into theater.

The briefings were grim, as the war was escalating, and we were essentially told to fight to the death. Otherwise, we would probably be captured and decapitated, and the video would be published on YouTube.

On one of the live fire ranges run by government contractors, I decided to adjust the sights on my squad automatic weapon to better adjust for sight picture from a standing position, and I was assaulted by a contract range instructor. While firing in the prone position, the instructor straddled my back, grabbed my helmet, and pulled up on my helmet to choke me with my helmet chin strap.

I maintained my composure and gave the contractor one warning to take his hands off me.

He didn't comply, so I grabbed him by the legs and flipped him forward onto his face, crashing into my loaded and ready-to-fire machine gun.

I used all my training and grappled my way out of harm, gained the advantage, and Harry stopped me from beating this man to death.

I have no clue what this man was thinking, but I suspect that he was prior military and believed that he had the right to act aggressively toward his customers.

The man was very lucky.

I knew that I had the right to kill him based on the rules of engagement, and I intended on doing so.

You probably can imagine the shitstorm this caused in my chain of command.

After Harry broke up the fight, it was a race to report the incident up the chain of command, with the contractors racing against the soldiers.

About five minutes later, I was explaining to the senior NCOs what had occurred with nearly fifty soldier witnesses present, and I plainly stated the facts. Shortly thereafter I was standing in front of the head contractor with the man that assaulted me and my senior leadership.

I did not back down and stated that I probably had the right to kill the man based on the rules of engagement, and before I took another breath the lead contractor was apologizing profusely and scowling at the man that had assaulted me.

The man that assaulted me suddenly realized the gravity of the situation, that the contract could be canceled for cause, and then apologized profusely. We shook hands and, as I walked out, the senior NCOs were smiling and laughing, patting me on the back as I walked away.

The man that assaulted me was lucky that the SAW didn't discharge and shoot him when he fell on it.

After that odd day, we were finally briefed on our mission.

* * *

We were being sent to Camp Ashraf, a forward operating base (FOB) in the northeast of the Suni Triangle located just north of a town named Al Khalis. At Camp Ashraf, our company's mission would be mostly typical for combat units during the height of the Iraq War. [38]

We conducted area security patrols, convoy security, ambushes, and cordon and searches; established checkpoints; and cleared military supply routes of enemies and improvised explosive devices (IEDs). There were also plenty of strange missions due to the nature of Camp Ashraf and the unpredictability of the upper command's needs during combat operations.

One of the strange things about Camp Ashraf was that it was the main military installation and home of People's Mujahadeen of Iran (PMOI). [39] The PMOI was designated as an international terrorist organization that was committed to overthrowing the Iranian government. The PMOI fought heavily alongside Saddam Hussein against the Mullahs during the bloody Iraq/Iran War.

During the US-led invasion of Iraq, the Allied forces heavily bombed and shelled PMOI camps in the southern part of the country. This caused the PMOI to strike an agreement with the United States to not fight against US forces in Iraq. Strangely, the PMOI terrorists were our allies against Iran. A part of this agreement resulted in the Allied Forces designating the PMOI as protected persons in Iraq in exchange for disarming and allowing the US military to protect the PMOI. All PMOI members were consolidated at Camp Ashraf.

If that wasn't strange enough, when the US occupation forces took over Camp Ashraf from the PMOI, many of the PMOI members defected to the United States.

The defecting members were placed into a temporary internment facility (TIF), which was like a small prison where the PMOI lived a strange existence contained to a facility but without the restrictions and structure of a prison.

These defectors were routinely interviewed by the CIA and other interested US government agencies to collect intelligence.

We would later learn that many of the PMOI defectors were not happy about being in the TIF, and routinely tried to escape, causing us to hunt the escapees down (because they were still technically terrorists, and we couldn't merely just allow them to go free).

Shockingly, these terrorists came from all over the world. One day, while working the entry control point to Camp Ashraf, a western white woman in her late fifties approached the gate from what appeared to be a rental car, handed us her Canadian identification, and asked if she could see her son who was either still in the PMOI or who had defected to the coalition forces and was in the TIF. Telling this woman that she couldn't see her son, but that he was in our custody, was surreal.

Over time, I started to realize that it is almost impossible to fight a terrorist or group's ideology with the military unless you kill most of the opposition and enact strict military law, like the United States did in Japan and Germany after World War II. Ironically, our base's mission was to protect both terrorist groups, in addition to our other combat responsibilities, and we were ready to get work in Iraq.

After a few weeks in Kuwait, we departed on Air Force C-130s to Iraq in full battle gear. None of us knew what to expect, and as we approached Balad Air Base (a.k.a Camp Anaconda)[40] and Logistical Support Area Anaconda, the aircraft began a steep downward spiraling maneuver toward the ground, and we were notified that the runway was receiving incoming mortar fire. My face and sinuses had sharp pain from the rapid decent.

Suddenly, the aircraft leveled off, pitched up, and landed hard and fast. The aircraft rapidly taxied and the rear gate was lowered as fast as possible so we could run and take shelter in bunkers near the taxiway. It was dusty, smoky, and we were disoriented.

That night we barely slept as fighter sorties were taking off next to our sleeping quarters and landing all night, coupled with the periodic incoming indirect fire alarm blaring with occasional mortar explosion off in the distance. I don't think anyone slept that night.

The next day we were loaded like cattle into armored truck beds and driven out to Camp Ashraf. We were given instructions to stay down and not look out of the trucks to avoid the enemy attacking a highly valuable target full of soldiers. Not a single person glimpsed over the side.

Then about an hour later, after what was a windy and bumpy ride, we arrived at Camp Ashraf.

Camp Ashraf also had an interesting feature: it was the home to the largest bunker complex and cache of weapons and munitions in Iraq. Saddam trusted the PMOI, and the PMOI were known for being ruthless, especially with Iraqis that were a threat to the PMOI or Saddam's regime. The PMOI were viewed as henchmen by the locals.

Nevertheless, they were entrusted with guarding this massive weapons cache before the United States had invaded.

The irony is that most of these weapons and munitions were provided to Saddam by the US government. The cache of weapons and munitions was so large that every day at 3 p.m., a twenty-five-ton explosion occurred to destroy the munitions. This daily demolition reportedly began in early 2004 and was continued when we left Camp Ashraf at the end of 2005.

One day while on patrol, three of my squad mates were nearly killed when a contract explosive ordinance disposal team working for KBR, Inc. forgot to communicate the blast to my squad mates that were sitting in armored trucks, called Highly Mobile Multi Wheeled Vehicles (HMMWVs; often pronounced like "hum-vee"), pulling security within the pressure wave blast radius. Luckily for them, they were inside the truck with the doors closed when the detonation occurred.

If they were a hundred meters closer to the blast, which is not a great distance in a vehicle, the shock wave could have killed them or caused serious permanent damage beyond their hearing loss. There were many ways you could get killed in Iraq, and many of them were due to negligence while performing dangerous work.

Thankfully, the army fired those contractors the next day.

Like numerous other combat veterans have reported from wars throughout history, war is 90 percent boredom. The other 10 percent is pure fear, adrenaline, and organized chaos. You could never predict what was going to happen or when these unknown events would occur and that is the terror of war.

Soldiers spend a lot of time staring out into the distance scanning the environment and watching everyone's behavior to identify bad actors living among the civilians with the goal of destroying threats before they destroy you.

We calculated that, as combat arms soldiers in Iraq, our chances of being killed were about one in twenty-one in 2005.

Luckily, nobody from my company was killed during our time in Iraq, but several people from our company were seriously wounded or injured. Our battalion was not so lucky, and numerous people from our battalion were killed or so seriously wounded that they were immediately medically evacuated to Walter Reed Hospital in Bethesda, Maryland; if they survived, they are typically either in unresponsive comas or living with severe disabilities.

Unless you personally knew one of the wounded, you often never knew what the outcome of the person was once they were medically evacuated.

I witnessed and participated in many terrible things in Iraq.

I also had a failure that is burned into my memory to this day. While on patrol, two civilian semi-trailers collided head on while driving down a two-lane highway at a high rate of speed. I was the person with the most medical training, and I began to triage the victims. I determined that one could not be saved and that one person with serious closed leg fractures and a head laceration only needed to be stabilized.

The last was in critical condition and could be saved.

I determined that he had severe internal bleeding in his abdomen and that he was suffering from hypovolemic shock due to blood loss. If I could get fluids into him, then we could buy him enough time to be medically evacuated and treated at the hospital.

I began to initiate an IV, and I could not hit the veins in his arms, as they were collapsing, and, as on time went on, I couldn't get the IV catheter into his veins. I must have stuck the poor man at least ten times, exhausting my supply of sterile needles before I had to reuse a dirty needle to try to save the man's life.

On my final attempt, I thought that I had succeeded in inserting the catheter to only notice a few minutes later a raised bubble under the man's skin at the insertion site, a tell-tale sign that the IV was infiltrating saline fluid between his skin and muscle tissues. I pulled the catheter, and I watched the light go out from his eyes as I held his arm firmly.

I looked back and shouted for an update from our squad leader on the Blackhawk Helicopter medical evacuation, and I was told that there were no resources available. This meant that the military command had determined that there were other higher priority missions for the medical evacuation units.

Consequently, we placed the man in a civilian ambulance where I am almost certain he died. Later that night, after our thirteen-hour patrol mission concluded, I learned that our battalion tactical operations center made a mistake when making the helicopter medical evacuation request, and there had been resources available to evacuate the man.

It was reassuring to know that we had such outstanding soldiers supporting us.

This was the hell that was war.

Anything, at any time, could go wrong, and we all sensed that we could fall victim. I often worried that help wouldn't be there when we needed it the most.

There were several gruesome realities of war that I experienced and prefer not to discuss in detail.

Some of the low points were shooting at erratically behaving vehicles (an indicator of a suicidal vehicle bomber), and responding to and placing the remains of contractors, civilians, and military members that had been incinerated by explosive devices into buckets or bags.

During a mission headed south toward Baghdad from Camp Ashraf, we were a few hundred feet away from a massive explosion that detonated in a market, and I vividly recall driving the lead vehicle into complete darkness caused by the dust and smoke to only catch glimpses of pools of blood in the blast area and civilians carrying their limbs. Perhaps it was a botched attack on our squad of vehicles as the detonation went off too early to strike my vehicle. Watching civilians being obliterated by enemy explosives and small arms fire was burned into my memory.

There were plenty of other awful experiences, as can be expected from a year-long tour in Iraq during the deadliest phase of the war.

Maybe someday I will elaborate further on my experience in that country.

But not today.

CHAPTER 5

The Sandbox Ends

Despite the hardships, war is like graduate school for the combat arms where your thinking and skills are honed.

There is nothing like evaluating our own performance shortcomings on the battlefield on one day, and then resolving those shortcomings in discussions and training before the next mission. In the combat theater, we were constantly tweaking and testing new skills or tactics. Although, all this sharpening comes at a cost to physical or emotional health for any sane person.

Halfway through my tour in Iraq, I learned that more than half of my company had been prescribed SSRIs like Paxil or sleeping medications to help the soldiers with the stress and to improve sleep.[41] [42] I was diagnosed with PTSD in theater and was prescribed Paxil, which I stopped taking after two weeks because of how it blunted my affect and changed by personality.

I was nervous and scared when I arrived in Iraq, numb while I was there, and terrified when the end of the deployment neared, as I feared that I would be gravely injured right before I would return home. The adage that it is better to get killed early in a combat tour, rather than later, was certainly accurate.

I came home cold and numb to the world, but the experience overall was positive in the sense that it was the most valuable learning experience of my life. I became more disciplined, tough, and wise to the world.

While deployed, I somehow managed to complete two correspondence courses in biology and psychology at SCSU. When it was my turn to rest on extended patrols, I took out my textbooks and read in the security of an armored truck, and one of my NCOs proctored my exams as required by the university. I also learned much about the true nature of humans and human behavior.

We are a ruthless species that will fight for centuries over ideology, beliefs, and culture. I also learned more about global geopolitics, psychological effectiveness of psychological operations, and the futility of intelligence in combat.

In Iraq, I quickly learned that nothing was as it appeared. Whether it was an IED hidden in the cavity of a dead donkey or a pile of trash, or the local Iraqi vendor attempting to sell you poisoned chewing tobacco or watermelons, any poor decision could result in your instant demise, or even worse, could kill everyone else around you.

As I experienced in Honduras, the true nature of our missions, as they related to the ultimate goals and the stated mission in Iraq, were often hidden from the people that had to execute them. In a war like the Iraq War, everything was political, even on the battlefield in our area of operation. Neighbors had axes to grind against each other, neighboring communities hated each other due to centuries of bad behavior often along sectarian divides, and the lines were blurred between organized crime, external state sponsored insurgents, or the ideologically driven terrorists.

By the end of the war, the only thing that was clear was that the only real mission for US forces that had any chance of succeeding was ensuring our own survival. The idea that Iraq would someday bridge sectarian and political divides to become a western-like democracy was an unachievable military mission.

The longer that the military was used to achieve this goal, the more that the military's use seemed to have the opposite effect.

Every day in Iraq, we would receive daily intelligence briefings before our routine missions, or before emergency ad hoc missions. These intelligence briefings often consisted of information that we determined to be useless. For example, if we served 364 days in Iraq, then 360 of those days' intelligence reports consisted of intelligence to be on the lookout (BOLO) for white Toyotas, trucks, or cars, that were suspected vehicle borne IEDs.

The problem is that probably 70 percent of the vehicles on the road in Iraq were white Toyotas. This was hardly useful information.

Another example was when we received intelligence that insurgents managed to steal an armored HMMWV from Camp Victory (this is where Saddam's palaces were predominantly located in Baghdad, a short distance from the Green Zone).

The next day we received an intelligence report to BOLO for a sand-colored HMMWV. How would my platoon be able to identify whether or not a sand-colored HMMWV was a threat from this intelligence report?

These examples were representative of all the intelligence I ever received while serving in the army. Perhaps others that served in the military had a different experience?

One particularly odd mission came when we were ordered to escort roughly one hundred of the PMOI defectors back to Iran. The mission was odd because it was the PMOI defectors, who had been trying to overthrow the Iranian government for the last twenty-five years, who were requesting to be handed off to the Iranians.

Some of the PMOI that we were bringing to Iran were not even Iranian, which was even more puzzling to us.

Perhaps there was another purpose for bringing these individuals to Iran? We were quite concerned, as anytime the US military moved or escorted busses around the country it was like shouting to the enemy "please attack us" as these were highly valuable soft targets, and the terrorists could care less who was in the bus. Often the person detonating the IED was a kid that was being paid by the insurgents or by foreign state-sponsored actors to do it.

The mission had several high-ranking military officials and people that I suspected were from the State Department or CIA that were handling the details of the handover. It was the first time that we were told that we were the priority for direct air support from F-16s and other fighter aircraft tasked to our mission. We could quickly obtain close ground support from the aircraft overhead by directly communicating with the pilots via our radios.

On the way to the border, I was tasked with driving my company commander and a colonel since I was viewed as being one of the best drivers available. I preferred to be a gunner or dismount security on convoy missions, but you don't get to pick and choose your assignments. I also took great pride in being asked or told to drive—the duty comes with great responsibility. When vehicle operations come under attack or call for action, the drivers of vehicles have the life of the crew in their hands and a good driver anticipates the best maneuver and acts appropriately without being told to do so and listens and attempts to always protect the crew.

During the drive to Iran, I was driving the lead vehicle, the most dangerous and important position in a convoy. Suddenly, a pack of wild dogs ran across the road in front of the vehicle in a tight corridor in an urban area with vehicles and other objects lining the road. I had assessed that it was a dangerous choke point and accelerated through the choke point and ran over the dogs. The vehicle was loud, everyone was wearing earplugs, and the officers sitting in the back seats had limited visibility, and it was dark with low ambient light.

Then, suddenly our gunner started laughing hysterically and ducked down into the cab and yelled, "That's a dead fucking dog!"

My company commander, who didn't spend much time off the base, looked petrified and asked me if I hit a dog, and then said, "That is low class."

I didn't say a word in response and then the colonel, who was an attorney, shrieked and whined, "Life is precious."

I maintained my professionalism but in reality, I could hardly stop myself from bursting out into laughter at this colonel pretending to be a soldier. When we arrived at the border, the company commander approached me to chew me out about the dead dog.

I stood at attention and then parade rest while I received the lecture. When I was given the opportunity to speak, I explained the tactical situation, the rationale for not stopping (could have been a diversion tactic to stop, attack, and disable the convoy), and then explained why my actions were the best choice.

Everyone stared at me, and the NCOs smiled at me.

They knew that I did the right thing.

This was the kind of skewed perspective that existed in the military for those that worked outside the wire versus those that rarely left the security of the FOB (about 70 percent of the people that served in Iraq). Getting lectured by the command for bad actions, only to have the command reverse their position and agree with the junior enlisted and NCOs, seemed to be a continual struggle. Many of the officers did not understand the realities on the battlefield.

I liked my company commander, and I think he was merely making the best out of what was a stupid war with unachievable objectives at the time.

Upon arrival at the border, we crossed into Iran according to our highly accurate GPS devices. We moved into positions to provide 360-degree security, while the lawyers, State Department, CIA, and miliary brass met with Iranian officials.

In an anticlimactic fashion, roughly fifty ex-PMOI walked off a charter bus, across the border, and into a building in Iran, and one western-appearing man walked from the same building and into American hands.

He was smiling, shaking hands, and hugging the US government personnel that we escorted to the border. It appeared that we just traded a hundred "former" terrorist prisoners for one captured CIA agent, and there was nothing wrong with that.

I was happy for this man.

Simultaneously, we were speculating how long would it take the Iranians to execute these former revolutionaries. From that point forward, I was always on the lookout for CIA agents operating in theater. The more that we watched closely during daily operations in Iraq, the more that we noticed and caught glimpses of them at work.

We were jealous of their ability to move quickly and blend in, as well as their comfortable clothes and air-conditioned armored civilian SUVs.

The more time went on during the tour, the more we missed the simple pleasures of civilian life.

In early November 2005, we received word that our replacements were going to start arriving in waves, and I was ecstatic to be leaving Iraq. Admittedly, we sandbagged our last week and, when we were requested to go on patrols or check hazards out, we would report to the TOC that we were accepting and executing the task requested, but we would really turn off our GPS and sit in the desert, smoke cigarettes, and drink sugar-free Red Bulls that were delivered weekly on a pallet to each platoon.

After a week of sandbagging, I received notice that I was being replaced, and I was to turn in my combat load of ammunition and grenades. After a few weeks of packing and selling items that I acquired in Iraq to our replacements, I was sent to Camp Anaconda where we were to wait for the remainder of the company and then return to Ft. Dix, New Jersey. We were told that we would be home by Christmas, and the army kept its word.

At Ft. Dix we sat through a week of debriefing and military out-process-ing (and drank very heavily) before we arrived at the Minneapolis Airport where we received a hero's welcome and state patrol escort back to the armory in St. Cloud, Minnesota.

People had draped flags on overpasses and had signs as they stood by the side of the road waving flags. I felt proud and wondered what they knew about the war. After a brief formation and speech, we were released to our families. I walked over to my father, and he hugged me and shook my hand for one of the first times in my life, and I hurried out of the building and into my father's car.

Home at last.

CHAPTER 6

What Is Wrong with Everyone?

Adjusting to life after back-to-back deployments was much more difficult than I anticipated.

I moved into an apartment with my girlfriend and transferred from SCSU to the University of Minnesota. In transferring from SCSU, I decided to switch majors from Finance and Economics to my new interests in science, research, and psychology. These types of major life changes and decisions are common after combat as the combat experiences change your values, beliefs, and perceptions.

I sought counseling at a United States Department of Veterans Affairs (VA) Vet Center for assistance obtaining my veterans benefits, as well as readjustment counseling services, and I ended up being offered a position as a program assistant at the Vet Center.

It was the perfect opportunity for a psychology student and provided a great salary and stability during that period of my life.

A few days after my arrival home, I noticed that I was aggressive and biased toward initiating fights and finishing them first. As they say, a good offense is a good defense. Everything and anyone could be used as an excuse on which to take out my anger.

I also had reoccurring nightmares which began halfway through my tour in Iraq, and which continued despite my best attempts to numb myself with alcohol. I decided to enter mental health counseling at the VA Medical Center in Minneapolis, and I started frequently running into other soldiers that I'd served with in Iraq or knew, and we often made jokes about our situation. Some of my friends were doing worse than I was at the time.

At a rate of one to two people per year, my friends that I served with have died due to drug overdoses, strange accidents, suicide, or while engaging in extremely dangerous activities.

One of the strangest deaths occurred when a friend and NCO that I served with in Iraq decided to carjack an Uber, and took the vehicle joy riding on Veterans Day, only to end up in police pursuit before crashing the vehicle at a high rate of speed for no apparent reason.

After I noticed the pattern forming after three years of strange deaths and suicides, I stopped attending the funerals.

I was eager to make up for lost time and decided that I wanted to become a psychologist. At that point, I was working full time at the Vet Center, attending classes full-time at the University of Minnesota, and was back to drilling in the Minnesota National Guard.

My first weekend drills were awful.

Nearly the entire 34th Infantry Division seemed as if it were deployed somewhere. Harry and I were placed in charge of new recruits and soldiers that had just completed infantry school. The new soldiers were eager to deploy to Iraq, and Harry and I just shook our heads and attempted to tell them to be careful about what they were wishing. (Harry hadn't escaped Iraq unscathed. At one point in the war, he was plastered with a White Phosphorus IED, which he was lucky enough to avoid hitting his exposed skin or face.) We attempted to tell them candidly about some of the horrors of war in the hope that it would make them think more rationally. It didn't work, and I am nearly certain that they probably found the war that they were seeking over the next few years.

That same drill weekend I met with my former company commander in Iraq along with one of our high-speed medics, and we were told that we were being recommended to attend Officer Candidate School (OCS), if we wanted to attend, and we both accepted the offer.

OCS in the National Guard can take a few different forms. You can attend monthly drills and be in OCS one weekend a month for eight months, and then attend an OCS school for three months during the summer to finish the program, or you can attend OCS with the active-duty force at Fort Benning, Georgia, which is a full-time commitment.

As having a college degree was a requirement, my only option was to attend the state program so that I could push quickly through school. OCS was to begin the following month, and I was excited to attend.

OCS kicked off with a week- or two-week-long extended training session, as I recall. We were provided with a packing list, and we were to go to our quartermasters and obtain everything on the list and report to the first training session. Typically, one of the first things that they do in the army is inspect your equipment.

So, I went to the supply sergeant's desk and handed him the packing list. But I was not able to obtain most of what was on the list as supply inventory had been drained due to the entire brigade being deployed to Iraq. After combat, we developed a laissez-faire attitude toward military process and customs, as it was set aside for combat.

In war if we needed a piece of equipment, or needed to modify our equipment, we obtained it or modified what we had by any means possible. We had learned that we could bend the army rules if it resulted in successful completion of our missions.

When I arrived at OCS, the OCS cadre were very upset that I didn't have everything on my packing list and that my equipment had all been modified. I also received counseling for bringing pharmaceuticals to treat my nightmares and anxiety, which had been prescribed to me by the army.

In the army, you can somewhat assess the type of person that you are dealing with by examining their gender, level of fitness, occupation, military schools that they have attended, and wars they have fought in by a quick examination of the badges and insignia worn on their uniform, as well as their physical appearance.

My instructors had mostly never been deployed, did not work in combat roles, and were not in great physical shape. Despite their shortcomings, I decided to push through what I knew would be a healthy dose of stupidity from unqualified instructors.

I wasn't disappointed. Only a few hours in, I was listening to the OCS cadre scream at a candidate using the horrors of war as a fear and motivational tool. The only problem is that this instructor had never deployed and the greatest hardship he had probably ever endured in the military was when the dining facility ran out of brownies.

It was extremely difficult for the medic and I to stay motivated in the training, and we often joked and commiserated with the misinformation about war and tactics that they preached to these green officer candidates.

At OCS, you take turns in various leadership roles, and this is common practice at many military schools. The other thing that can be expected is that the instructors will change the plans multiple times and give you repeated obstacles. The idea is to see how you will perform under high stress as a leader, and to see how you react to failure.

A few days into training, I was given the pretend rank of battalion commander. This is typically a lieutenant colonel. A battalion commander's role is to manage and lead, usually four or five companies, with a total of roughly 750–1000 soldiers under his or her command. The night before I assumed command, I had to prepare an operation order for the battalion, and then lead the field training exercise and movement to and from the training area.

I quickly assessed the operation order and knew it was likely not achievable, as expected. So, I went to the OCS cadre that was acting as my brigade commander and made my case to request changes to certain aspects of the operation order, which to my surprise was granted.

The next day we traveled to the training area, and everything was going according to plan, and I was waiting for them to start changing things and playing games.

We made it through lunch and there were no hiccups.

After lunch, we had to move to a different training area via a combat march, a tactical way to travel distance by foot, a few miles or less.

That's when they started issuing changes, or fragmentation orders, so I went to the company commanders and reiterated that we had to accomplish the mission by any means and gave the company commanders a ranked priority list of the objectives, as some of them were not necessary to accomplish the mission.

For an hour, the cadre couldn't figure out what exactly I had done, but they were upset. Eventually, they figured out that I eliminated a few objectives from the operation order and the screaming began. They attempted to chew out one of the company commanders, and I immediately claimed responsibility and told the OCS cadre what I had done.

They were livid.

They started screaming at me for not following orders, and I calmly stated that I had followed the commander's intent and accomplished the mission.

Did it really matter where or how the companies hung out their sleeping bags to dry?

No, and commanders on the battlefield don't waste their time micromanaging these types of decisions, as it is a waste of time and energy.

When I explained this, the cadre lost their minds. I was ordered to low crawl—the most difficult type of crawling, that is performed by dragging and not lifting any part of your body off the ground—for half of a kilometer, and I smiled and laughed the entire way.

***

After the first week session of OCS training, we were released back to our civilian lives.

My manager at the VA, "Dr. Bert," was not too thrilled about me missing time at the office to attend military training.

I found this quite odd as he was a Vietnam veteran himself. The other staff members in the office were mostly combat veterans, or at least had prior service, except for a few counselors that simply loved helping veterans.

While working at the Vet Center, I was trained as a contracts technical representative, sometimes called a CO-TAR in the US government, where I learned about all the rules, laws, and regulations of government contracting work.[43] At the time, the VA was looking to rapidly expand services geographically to serve the massive wave of veterans returning home from overseas.

I had the responsibility of identifying areas where the VA needed to provide additional services based on veteran density and accessibility to existing offices and services and had to identify new potential locations or healthcare providers that could provide services to veterans via contracts. All these tasks required close collaboration with the contracting officers, which were mostly lawyers or seasoned government employees, and I learned quite a bit about government contracting and procurement.

While it didn't seem important at the time, the training and experience was invaluable throughout my career.

While working at the Vet Center and studying psychology, I made the decision to become a psychologist. At school, I was loving the work and the University of Minnesota psychology program was one of the best research programs in the world. The undergraduate students had different track options.

One track was research-focused, which required a research thesis and had an emphasis on science and statistics courses, and the other track required that the student complete a capstone project which was more akin to the liberal arts. Out of the five hundred psychology students in my graduating class, there were only forty of us on the research track and we all knew each other.

As part of my studies and on the path to become a psychologist, it was recommended to start internships supervised by a licensed psychologist and by a member of the faculty. I approached Dr. Bert and requested to intern with him during what would be my junior year of school, to which he agreed.

It was bad enough sitting through my own therapy, let alone watching someone else conduct therapy. As a psychology student, you can sit and pick apart what the counselor is doing and respond to the techniques they are using in therapy.

When psychologists see other psychologists, it's not like what the rest of the population experiences.

Sometimes, full-blown scientific debates will take place related to the patients, perceptions, cognitive processes, and behaviors. During the internship, I discovered that I hated counseling patients, and I had a terrible bedside manner. Whenever they would present a problem that they thought was significant, I couldn't escape framing their concerns within the context of what I had experienced or known about the world.

I knew then that I would never become a psychologist.

To become a psychologist, you must perform at least one to two years of clinical work, and I had no desire to counsel people. I only wanted to study them. That presented a big problem as I didn't know what I was going to do after school.

Despite all these uncertain things happening in my life, I was merely staying busy to avoid my own PTSD. The next month before OCS I hit a tipping point and began to have multiple panic attacks. I had begun experiencing them a few days after I returned home from Iraq, but they were increasing in frequency and intensity to the point where I could take a few milligrams of clonazepam and antihistamines and it would not stop them from occurring.

My body and mind still felt as if I was at war.

I was doing well in school, work, and in the military, and it all crashed after a few months home. My startle response was heightened, I couldn't concentrate, and I was losing motivation to do anything that I enjoyed before the war.

Before the next OCS drill, I received a phone call from my readiness NCO, and I received notice that I was being promoted to the rank of sergeant (E5), if I wanted to drop out of OCS. It was perfect timing.

Even though my PTSD was making me anxious, angry, and agitated, I decided to go through one more drill to see if the training would improve. Unfortunately, it did not improve, and on the Sunday of that drill, I made two decisions: the first was to quit OCS and take the promotion, and the second was to not reenlist at the end of my tour of duty, which was rapidly approaching.

I requested to meet with the OCS officer in charge on Sunday night, and frankly told him about how this OCS program was failing the army. I could tell that he was insulted and did not care the slightest about what I had to say. I also attempted to speak to him about the retention problems in the army due to numerous failed policies, and he flat out told me that he didn't care about that either.

I then informed him that I was quitting OCS and he instantly turned bright red: his highest decorated officer candidate, with strong battlefield recommendations, had just quit.

I saluted, did an about face, and walked out of his office.

I was happy to be leaving the National Guard within the next year.

***

While I was simultaneously working for the VA and serving in the National Guard, the command began asking for my assistance to help other men from the unit with various problems.

One of the soldiers lost a leg in an accident, and the unit requested that I help him navigate the VA system or with any other issues that he was having. I graciously obliged.

A different soldier that I was friends with walked into my office at the VA and said that he needed to see someone, and I went back to Dr. Bert's office and asked if he had the time to see an unscheduled patient. He had the time, so I walked with my friend back to his office and introduced him to Dr. Bert.

I went to my desk and did paperwork for the next hour, and my friend stopped by my office, thanked me, said goodbye, and walked out the door.

A few minutes after he left, Dr. Bert walked into my office, shut the door, and asked me if I knew that he had a heroin problem.

I was shocked and said I really had no idea.

A few months later, he died of an overdose.

While working at the VA in mental health care, I had greater insight into the soldiers in my former Iraq unit's problems and greater insight into the combat veteran community at large; sometimes it may be better to not know the truth about what is happening around you.

While working for the VA, I successfully relocated our office; assisted with the acquisition, contracting, and opening of two other new facilities;

trained my counterparts at the new offices; and consequently, I received a government bonus and award for my performance as a civil servant.

I took pride in being a government employee and did my best to save taxpayer money whenever possible while ensuring the best possible service to our clients and taxpayers. My supervisor, Dr. Bert, was hardly an effective leader. He played politics, was dishonest, and routinely played games with staff members or treated them unfairly. He also didn't know how to use any of the government computer of software programs to manage clinical records, manage the budget, or how to perform a long list of administrative tasks.

That means that I was responsible for all these tasks. I didn't mind doing them except for the fact that Dr. Bert would then attempt to micromanage things that he didn't understand when it was convenient for him.

I also learned that Dr. Bert had sabotaged a promotion that I was seeking which would have paid me an additional $25,000 per year without cause. I presume it was so that he would still have me doing his work for him.

He was diagnosed with cancer, and he was out on medical leave for about a year. That enabled me to at least keep doing the same work without his constant micromanagement and terrible leadership.

Without him acting as an arbiter, I had direct contact to regional management and was able to communicate with VA senior leadership in Washington, DC. This gave me the opportunity to learn about the nuances of government spending including cost centers and funds, loopholes, and the legal provisions to move money across accounts to get our work done.

VA Secretary James Peake even stopped by my office in St. Paul, Minnesota to thank me and gather intelligence about how his mission was being executed at the ground level. He was surprised that I was so young (twenty-six) to be in my role but was less surprised after I told him that I too was an infantryman.

The inside joke in the army is that infantryman believe that they can do anything. Secretary Peake had a long and distinguished career in the army where he first served as an infantry lieutenant in the Vietnam War.

After I'd successfully managed our office and several large projects, Dr. Bert was beginning to recover from his chemotherapy treatment and started coming back to the office periodically. For a while, he was a nicer person at work after his brush with death.

However, his behavior slowly returned to normal and eventually he went back to micromanaging me.

One of my other duties was that I was the network administrator for our office. The previous program assistant was tech savvy too and he'd begun to digitize the old paper clinical records. I completed the task after his departure.

I also created a backup system for all the computers in the network, and while performing routine maintenance on the backup server, I discovered that Dr. Bert had been running his private counseling practice in the government office.

This is a serious crime and was essentially government theft, so I reported the crime to the Regional Office and the VA Office of the Inspector General (OIG) when I discovered it.

A few weeks later, the regional manager arrived at my office to take a statement from me and took the server as evidence. I didn't suspect that anything would come it. A few weeks later, we moved into a new office that I had procured, which was a significant and much-needed upgrade from our old dingy, dark, and dank-smelling office building. We finally had a pleasant environment for our employees and veterans.

A few weeks after we arrived at the new office, Dr. Bert made an announcement that he was retiring. Dr. Bert never spoke to me once during this period, which was uncharacteristic of his constant attention-seeking behavior.

I finally caught wind from other VA employees that he was forced into early retirement over the theft from the government. Helping show that man to the door was one of the best things I had ever done for taxpayers and the government.

At that time, I knew that I had to focus on my next career steps, and I quit my position at the VA.

***

Between 2006 and 2009, I had married my longtime girlfriend, and the marriage later came to an end in the blink of an eye like so many other military relationships. Thankfully, my PTSD was better managed thanks to the help of Dr. Thad Strom[44] and Dr. Eric Brown from the Minneapolis Veterans Affairs Medical Center even though my occupational impairment from the combination of my injuries and disorders was increasing.

I was in my senior year at the University of Minnesota, and I had managed to maintain a 3.5 GPA through my most difficult courses in study design and statistics, and in core psychological research courses. I almost had "failed" biological psychology, which was a course that often derailed many pre-med and doctoral students in their senior year, by only earning a B.

At that point, I still didn't know what I was going to do next.

I worked as a research assistant in several of the best laboratories in psychology, and I learned more from working in these laboratories than I did from all my course work. I learned advanced statistical methods, hand keyed and memorized numerous psychological tests and instruments, and learned how to design and implement complex research protocols in occupational interests, personality, behavioral genetics, clinical, and counseling psychology.

I had a burning passion for all things science and befriended many of the faculty across campus and departments. I would often spend my free time hanging out with old expert research technicians asking them questions.

My senior year, I took the GRE and applied to PhD programs in psychology at some of the best schools, despite not wanting to engage with clinical work.

I thought about going to law school, took the LSAT, and then landed a judicial externship in the Minnesota Court system without attending a single day of law school, likely due to nepotism (I had a personal relationship with one of the judges which enabled me to get my foot in the door).

I was fascinated by the law and court, and I had spent many times as a defendant in traffic court and for a long list of mischievous minor offenses prior to my military service. While working in the courts, I learned about researching case law, how to argue and write persuasively from any perspective, and how to logically break problems apart while focusing on the facts of each case.

Every morning began with all the judges having coffee and tea together with the externs, where the judges that clearly had different political and personal beliefs all discussed and debated various legal issues and current events. I quickly ascertained that these judges were all incredible people doing a great service to our community. Many of them had left lucrative careers in industry or private practice to serve as judges, and their jobs were not easy.

They often had complex cases to deal with which required hours of reading and writing that often consumed their nights and weekends. I became friends with two of the judges and we often ate lunch together. They mentored me and would give me hypothetical cases or court objections that I was to rule on to build my legal acumen.

Through these critical thinking legal drills, I began to understand how to identify the critical issues in cases, and how to be impartial. I spent hours writing decisions for the supervising judges, only to turn in my written decision and then be told to write the opposite decision.

At first, I found this frustrating, but I quickly understood what the judges were doing—they were teaching me how to quickly identify counter arguments and address them in my writing.

During the six-month externship, I had several mini assignments. For the first assignment, I was loaned to the appeals court for a few days, which was a more formal environment where I learned the basics of why cases were often appealed and how cases became more political at that stage (one of the appellate judges was a family friend from a lesbian power couple). Sometimes, lower court judges would make simple process errors which would result in the cases being sent back to lower courts, and less often there were cases that were challenging specific aspects of the legal process or the constitutionality of law.

The second assignment was with a petitioner's attorney whose practice primarily focused on injury lawsuits against insurance companies or organizations with deep pockets. Interestingly, I learned that his main job was to lose cases and that it was a numbers game. By losing the most cases possible, while trying his hardest, he survived on the good word from his clients. Even if he lost the case, but his clients felt like he had tried hard to win their case, they would come back to him if they ever needed him again, and they would refer their friends and family to the firm.

It was a brilliant strategy.

Lastly, I spent a week at the coroner's office, where I worked with the head and assistant coroners. I learned about collecting trace evidence, performing autopsies, and determining the cause of death in the context of the forensic evidence.

At the end, I was given an exam consisting of only photographs of twenty homicide scenes and, apparently, I was the first person to accurately determine how the people died. The strangest was a man that committed suicide with a knife by carving and cutting into his own heart.

I determined that the small cuts were superficial test cuts, which, as the man became more desensitized to the cuts, he initiated longer and deeper cuts, until he finished the job. At the end, they offered me a position at the coroner's office, which I politely declined.

During the externship, I learned that most cases are quite like other previously adjudicated cases, and the judges spend most of their day corralling the lawyers and rewriting similar cases, repeatedly.

I eventually found legal work to be boring and no longer felt like a career as a lawyer matched my interests. But I thoroughly enjoyed my time clerking and the professionalism in the courts and gained much respect for the system after having worked in it. But I knew I had found my true passion.

* * *

As graduation rapidly approached, I learned that I had been accepted to a few of the programs where I'd applied across disciplines.

The only problem was that my wife and I had recently purchased a home, the housing market had crashed, and we were less than thrilled about trying to sell the house and relocate to different states.

As I was mulling over what to do next, I caught wind of a new degree program being offered at the College of Science and Engineering at the University of Minnesota called a Master of Science in Security Technologies, and I decided to attend a listening session presented by Professor Massoud Amin.

At the session, I learned that the program was a national security–focused engineering program where the students could focus in any of the core areas of national security using various technological or engineering approaches. It was a very applied program, and, with my military experience and security clearance, I thought that I could easily carve out a niche for myself in national security.

Just before graduation, I applied to and was accepted into this new program, with my courses beginning about two months after graduation.

CHAPTER 7

Thinking Security

After a much-needed break after graduation, I began my master's degree classes during the summer of 2010.

Graduate school was exactly like I expected since I spent quite a bit of time palling around with graduate students as an undergrad.

My wife completed her master's degree and accepted a dream position in her field on the West Coast. This was the beginning of the end of our relationship.

We obtained a property near her new office, and we moved her belongings from Minnesota to the mountains in the California. The plan at the time was that I was going to complete my master's degree as fast as possible, and then obtain a position in corporate security in California or Nevada.

While visiting my wife only a few months after I had left, I discovered evidence that she had been cheating on me. I confronted her about it, and, given the circumstances, I would have forgiven her if she could have been open and honest about what happened, but she chose to be dishonest.

I gave her plenty of chances to come clean, but eventually I had to end the relationship.

I could not handle being married to someone who was untrustworthy, undependable, or dishonorable, which were my least favorite qualities in others. The split was mostly amicable and easy as we had already physically separated, had split our belongings, and did not have any children.

A person who I met in my graduate program jokingly referred to it as a practice marriage, and he was right.

I learned how to properly behave in a serious relationship with a woman, and, to my ex-wife's credit, she dealt with a lot of anger and hostility from me at the height of my PTSD.

* * *

The expected reading load in graduate school was ridiculous and typically unachievable for a normal human, so learning how to skim and identify important tidbits from the assigned readings was paramount to achieving good grades. In graduate school, you are only required to take six credits to be considered a full-time student, but I took no less than twelve credits per semester.

My new graduate class had roughly thirty-five students, of which only four were women. This was in stark contrast to my undergraduate psychology program which was 90 percent female. There was a wide distribution in age, with the youngest student being twenty-two and the oldest student being nearly sixty years old.

This new security-focused graduate program used a cohort model where teams consisting of four to five graduate students were established on the first day of class and would persist across the program's core courses and throughout the duration of the program. I was very lucky as my teammates were rockstars.

Among us, there was a chief security officer of a medium-sized regional bank, a seasoned logistics manager, a cyber and information security executive, a whiz social scientist, and only one team member that was dead weight. Many of the other teams only had one or two strong students, and often the burden of the workload fell on those students.

I'd hated group work throughout my academic career for that very reason, as I was always the student doing all the work. By the time I was a senior, I typically did the project myself unless there was another student that would contribute work that was not garbage.

Doing the entire assignment, yourself, is much easier and faster than trying to clean up someone else's incoherent mess.

Typically, each course consisted of weekly presentations for each group and weekly individual written assignments. For the students that had been out of school for a while, it took them a while to adjust back into the grind of completing reading and writing assignments every few days. I understood why graduate school only required six credits, as many of the graduate students across the university had full-time research assistantships or were working full time as professionals.

I did not mind picking up my teammates' slack on occasion as they had much wisdom and insight into their areas of expertise and were well connected in the professional world. As one of these classmates frequently said, "You don't go to graduate school for the coursework; you go to graduate school to grow your professional network for your next career move." He was right: a person does not need to go to graduate school or college to learn a subject.

You only need to be driven, curious, and passionate to become a subject matter expert.

The Master of Science in Security Technologies program focused on teaching students how to be leaders and technical experts in any of the sixteen sectors of national security critical infrastructures, including cyber, physical, and virtual security systems.

The program used a holistic approach to train the next generation of security leaders with the critical skills and foresight needed to address real-world security challenges. The program was specifically designed to teach security professionals to understand and design resilient systems to identify and mitigate threats, vulnerabilities, and risks related to the inter-dependencies between critical infrastructures.

Each student was able to tweak the program to fit their career goals in combination with

electives. As we would learn, any security risk was multi-faceted and often multidisciplinary. Conducting risk, vulnerability, and threat assessments typically was more effective when there was a diverse group of experiences and disciplines in the room. Groupthink results in a lack of imagination, unforeseen vulnerabilities and threats, and ultimately poor security assessments.

When fall courses arrived, I upped my credit load to fifteen to eighteen credits per semester, which is considered an insane amount of graduate coursework. I took it in stride, and I decided to minor in geographic information systems (GIS).

What is GIS?

A geographic information system (GIS) is a computer system that analyzes and displays geographically referenced information. It uses data that is attached to a unique location. I had an elective with a famous GIS scientist, Professor Shashi Shekar.

Prof. Shekar is a legend in the GIS community, and I immediately saw the power and value of quantitative analysis using GIS. The GIS program at the University of Minnesota was taught by incredible instructors, and I learned how to use super computers for spatial analysis and statistics, the principles of cartography, the laws of geography, and how to write spatial analysis code in SQL, C++, and CUDA.

I also learned how to perform statistics with a software program called R. CUDA was hot in 2010 because NVIDIA had released GPU technology that made super-computing achievable on a computer workstation, and, later in the program, I completely fried the circuit boards on three new NVIDIA GPU boards by running a complex spatial analysis and simulation on them. If I would have liquid cooled the machine, the new computer that I built might have survived the ten-day simulation.

You live and learn.

Being a combat veteran in a national security–focused program was a significant advantage. I understood the tenets of physical security, had conducted countless threat and risk assessments, and understood what constituted perceived risk versus real risks, a common problem in the security world.

For example, I have heard countless times something like, "What if someone fired a rocket-propelled grenade (RPG) at an airplane and killed everyone as it was taking off?"

First, where would a person obtain a rocket-propelled grenade in the United States? They are not used by the US military, so it would have to be imported and smuggled in. Not an impossible task, but it was not exactly easy to obtain an RPG or simple to execute an RPG attack on aircraft during take-off either.

Second, even if you were able to obtain an RPG, then good luck hitting a speeding aircraft with it, especially on your first and only attempt. Hitting a moving object with an RPG is almost impossible, even for the best-trained RPG attacker.

While many of the students were spending a lot of time trying to understand how to assess and measure various security vulnerabilities and threats, the military veterans were focused on mitigating and triaging risks.

For this new program, we were instructed by some of the best professors at the University of Minnesota and leaders from the US government and national security industry. For example, Jeff Bender, Shashi Shekar, Shaun Kennedy, Mike Osterholm, Ken Kasprisin, Elizabeth Amin, and Michael Rozin where just a few of the core faculty.

We had guest instructors, including a lead scientist from the Defense Advanced Research Projects Agency (DARPA).[45] He developed exploding microscopic mechanical (MEMS) circuits that could coat a target with an invisible cloud of explosive. A different guest instructor was a former CIA agent that ran operations in West Germany during the 1980s during the height of the Cold War. We also had several guest instructors that were senior executive staffers working for the White House during President Obama's first term in office. The numerous connections that we made during the program made it almost impossible to not land a career in security.

While sixteen critical infrastructures would be a lot of material to cover, if you learn how to compartmentalize risks to individual facilities or system components, then you can mitigate the catastrophic risks, which can cause systemwide failure.

Isolation and compartmentation are key to all the security domains: security and risk management, asset security, security architecture and engineering, communications security, identity and access management, security assessment and testing, security operations, and incorporating security into design.

Security can never be 100 percent effective.

Security is like Swiss cheese; you must layer it to plug all the holes. So, if you can isolate and compartmentalize the risk, you can prevent cascading failures, or, in other words, a complete unmitigated disaster.

From a national security perspective, catastrophic risks related to weapons of mass destruction, like chemical, biological, radiological, and nuclear threats have been, are some of the most feared and well-funded since 9/11. I say feared, because I have learned over time—from being one of the few people that has quantitatively analyzed how security managers assess security risks—that often security managers tend to have biases that impact how they assess and mitigate risks, which is a significant problem in both corporate and government security.

So, I started to explore ways to make risk assessment more objective.

Based on my newfound love for GIS, and my realization that it was an underutilized tool in national security, I decided to focus on this problem for my master's thesis. I decided to see if I could spatially model and predict terrorism and violent crime.

I obtained a copy of the FBI's National Neighborhood Crime Study[46], analyzed this large data set, and combined it with the Census Bureau's spatial and population demographic data to create spatial rates of violent crime and terrorism. Then, I ran global spatial statistics and a spatial Monte Carlo analysis, a method of generating thousands of randomly generated points to statistically compare against, to validate my analysis.

What I found was striking.

The location of violent crimes and some forms of terrorism were highly predictable.

I completed the program at a rapid pace. My master's was completed in three months and the entire degree was completed in fourteen months (forty-one credits). After performing well on my master's thesis defense, my graduate committee strongly pressured me to obtain a PhD in public health.

Dr. Masoud Amin requested that I meet with Professor Jeff Bender from the Veterinary Medical School and the School of Public Health, and Shaun Kennedy, the center director from the National Center for Food Protection and Defense (NCFPD was renamed to the Food Protection and Defense Institute)[47], a US Department of Homeland Security Center of Excellence.

Dr. Bender had experience in agroterrorism, as well as bioterrorism related to food production and the weaponization and use of biological agents as weapons, and Shaun was a former food industry executive and engineer with expertise in food processing systems and supply chains.

At the time, I was tired of being in school, and I really wanted to begin a new career. At the meeting with Dr. Bender, he walked me through the program, stated that they had already identified a research thesis for me to work on, and shared that the data had already been collected (having the data collected knocks one to two years off the time it takes to complete a PhD). He also told me that the project came with a full salary and full scholarship; these types of arrangements are the ultimate offer that you can receive as a PhD student.

After we discussed these issues, I asked him which PhD program I would be completing, as I was uncertain. He thought that either the epidemiology or the environmental health programs would be the best fit for a career in bioterrorism or biosecurity.

I later learned that there were several other PhD students that were in the Environmental Health program at the University of Minnesota, most of whom were in the emerging infectious disease specialty track. Dr. Bender said that my work was very strong and that I would be admitted to the program immediately, if I chose to accept the offer, which I immediately did.

I was so happy and excited that, when I left Dr. Bender's office on the St. Paul campus, I was in tears.

A few days later, I met with Shaun Kennedy. Shaun was an oddity in academia as he had not earned a doctoral degree but was certainly worthy of the title as he had obtained enough experience in industry and skills to be worthy of it.

During our meeting, he explained to me the problem that I was to investigate for my PhD thesis. NCFPD, in partnership with the US Department of Homeland Security, had collected data of all potential threats and hazards on most of the food and agriculture systems in the United States.

At the time, this sounded like a straightforward project, and I was excited to start working at the center. We negotiated the salary, my title, and work expectations. I was hired as a Research Fellow due to my previous national security experience, which enabled the center to pay me at a much higher rate than what was typically offered to PhD students at the university. And since I already had the education benefits to cover my tuition from the Department of Veterans Affairs, I was able to further negotiate my salary higher as I did not need the scholarship from the university.

This was a dream opportunity.

* * *

I began working at the center a few weeks after meeting with Shaun, and I also met my co-advisor, Professor Craig Hedberg.

Craig was an expert in food contamination and food safety and had great insight into how food contamination occurs and how it spreads through a food system to consumers. At the center, I was first introduced to a fellow from the Department of Homeland Security, Bill Krueger, and Col. John Hoffman (retired).

Bill was a former laboratory director for the State of Minnesota.

Bill understood the vital role that laboratories play in the detection of harmful agents in food products or people, and how the laboratory system collected and analyzed samples from people, animals, and food at laboratories and how this information is communicated and shared across a network of laboratories to the federal government.

Col. John Hoffman was equally passionate about his work and is a highly driven individual. Col. Hoffman had a prestigious career in the US Army, and he helped establish the biological division of the US Department of Homeland Security, after it was created after 9/11. Like many people working in national security, he is highly passionate about his work protecting the US food supply, biological threats, and threat mitigation, and is full of energy.

Since I had changed disciplines for all my degrees, I had created more work for myself since I had to take the most basic courses to meet the requirements for each degree.

My graduate coursework for my PhD specialty track mostly consisted of coursework in advanced research methods, infectious disease ecology, infectious disease emergence, transmission dynamics, toxicology, epidemiology, biosurveillance, infectious disease policy, and biostatistics. For every course assignment where I had the opportunity to choose the subject matter, I always focused on chemical or biological agents that could be used in biowarfare or bioterrorism.

At the time, I was tired of coursework and viewed it as somewhat of a distraction to completing my dissertation research at NCFPD. Still, I had completed an additional seventy-one graduate school credits by the end of my PhD in only two and a half years.

At the end of my master's degree, I learned a painful lesson—you should immediately begin research and thesis work before your first day of classes.

I had waited too long to begin my master's thesis and almost put myself at risk of not completing my thesis in time for my self-imposed deadline. Based on that previous experience, I began my dissertation research immediately. I asked for a copy of the data that I was analyzing, the data collected via a software application named the Food and Agriculture Systems Criticality Assessment Tool (FASCAT).[48]

After obtaining the data and examining how it was structured and how the software was used in context, I identified a process to validate, verify, and evaluate the effectiveness of the platform. In context, FASCAT was used by state government officials across the country to collect information from private industry critical infrastructure owners in food and agriculture to determine which food systems or facilities were the most in need of being protected based on the economics of the food system evaluated and the ability for that food system to be used as a Weapon of Mass Destruction (WMD) delivery vehicle to kill tens or hundreds of thousands of people.

While working at NCFPD (a.k.a. FPDI), I was introduced to and worked with Dr. Amy Kircher. Dr. Kircher had completed her doctorate at a great program at the University of North Carolina and had previously worked for the Department of Defense as an epidemiologist at the North American Aerospace Defense Command (NORAD). It was my understanding that she was intimately involved in the public health preparedness planning and disaster response for all issues of public health in North America.

I would often look at the pictures of her shaking hands with President George W. Bush in awe. She was highly organized, professional, and understood every aspect of how the US government's public health system operated related to bioterrorism, biowarfare, and pandemic preparedness and response.

From 2011 to 2014, the risk management framework for critical infrastructures was a work in process. At the time, the US Department of Homeland Security (DHS) was asking the states to provide risk information on all hazards to critical infrastructures. In my opinion, one of the biggest failures in the creation of the DHS was the focus on all hazards.

This definition was so broad that anything that could be imagined as a hazard or threat to critical infrastructures could fall under the responsibility of DHS to protect. DHS was responsible for hurricanes, terrorism, cyber threats, floods, financial systems, earthquakes, military manufacturing, dams, forest fires, or even a person with a high-powered rifle shooting an electrical transformer.

The list of critical infrastructures essentially is a description of everything in a modern-day society. As the lockdowns during the COVID-19 pandemic proved, everything in society is critical and essential. Believing that the government knows what is best for all of us, and truly knows what is critical and what is not critical, is highly subjective.

This is what I successfully demonstrated with rigorous quantitative and qualitative research.

As I completed each chapter of my dissertation, I had it published in peer-reviewed literature. This is the best approach for scientists to take, as it prevents PhD committee members from executing political attacks against the student, which sometimes happens. It's hard for a committee to not graduate a student when all your work has been externally reviewed by experts and then published.

During my research, I was provided with what seemed like an almost unlimited travel and research budget. I traveled to several states where I observed how FASCAT was being used to collect data, held meetings with experts in the food industry and government to determine how the methods could be improved, and conducted every statistical experiment that was possible with the FASCAT data.

During this process, I greatly expanded my professional network with state and federal agency leadership across the country.

As my thoughts and interpretation of the research solidified, Shaun, Amy, and John began taking me with them on trips across the country to

meet with corporate leadership at some of the world's largest food companies. Additionally, we also frequently traveled to Washington, DC, to meet
with numerous stakeholders and government agencies.

Typically, over a period of two to three days, we would meet with eight
to ten people per day or attend day-long meetings where a government
agency or a food industry association would sponsor the meetings. It was in
these meetings I learned one of the most valuable skills as a scientist—how
to obtain funding for research from project sponsors and how to help influence the research priorities of the federal government and nongovernmental
organizations.

Obtaining research funding is much easier if the agency's leadership
knows who you are, if and when they like you, and when you are selling
their own, often crackpot, ideas back to them. The goal was always to identify which funding requests were coming in advance, prepare a menu of
concepts to later present to the stakeholders, and then see which concepts
the project sponsors liked most.

This process greatly increases the success rate of funding.

At the time, one of the best places to network with industry and government leaders in national security was at the various Sector Coordinating
and Government Coordinating Councils (SCC/GCC) for the various DHS
critical infrastructures.[49] [50] Currently, there are sixteen critical infrastructures which include almost everything that you can imagine: chemical
production, communications, dams, emergency services, financial intu-
itions, government facilities, information technology, transportation systems, commercial facilities, critical manufacturing, defense industrial base,
energy, food and agriculture, health care and public health, nuclear reactors
and waste, and water and water treatment systems.

The real problem the US government had when creating DHS was
determining what was critical to sustaining the functioning of our society. The answer was painfully obvious: everything in our society is critical
and the critical infrastructure list represents this. As the federal and state
governments in the United States pushed for lockdowns in March 2020,
fights began to erupt over which specific businesses and organizations were
necessary and critical.

As the domestic economy was crushed from the never-ending lockdowns, the answer became clear to everyone.

As a PhD student, I routinely attended the SCC/GCC meetings for
Health and Human Services and for Food and Agriculture. The meetings

were mostly to show off the work that was accomplished by the stakeholders that were receiving federal funding from one of the agency stakeholders sitting at the table. I later determined that the meetings themselves were useless, and as an independent scientist, the leaders of the industry associations often stated to me privately that they did not see much value in the meetings themselves.

The most important part of these meetings was what happened outside the official rules of the meeting, which bar pandering or soliciting. For the private industry leaders attending, they would collect intelligence and gain insight on what the federal government was planning. With that information they then had the ability to influence the government before they made rules or established policy.

The scientists attending these meetings were supposed to be there to speak truth to power, but it was much easier and better for your career to listen carefully and to grease the political wheels to obtain funding.

For the government employees, I think they were merely checking the box on conducting the meeting.

At these quarterly meetings, I routinely presented my work, and I also met numerous federal and state employees working in law enforcement and US government scientists working with Federally Funded Research and Development Centers (FFRDCs)[51], many of whom I am still in contact with professionally to this day. FFRDCs are operated by universities and corporations to fulfill certain long-term needs of the government which cannot be met as effectively by existing in-house or contractor resources.

FFRDCs were created during World War II to develop nuclear weapons as part of the Manhattan Project. After WWII, and through the Cold War, the number of FFRDCs expanded to forty-two institutions working on the cutting edge of a wide variety of problems related to national security and defense. At the time, I viewed working at one of these institutions as the paramount achievement for a newly graduating scientist.

I will never forget the first presentation to the DHS Food and Agriculture SCC/GCC at one of their quarterly meetings. I prefaced my talk with an image of the formula for a mixed linear regression model and the entire audience gasped when hit with the visual representation. I had not planned to discuss the formula; it was merely used to demonstrate my expertise.

After moving on to the next slide, I described how I had taken the hot mess of all the food and agriculture system threat and risk data provided to me by DHS and made sense of it on a colorful scatter plot, which demonstrated how DHS could generally and simply predict risk for each type of food facility or food production system in the entire United States.

I recall looking around the room, and I could sense that many people were impressed to finally see something meaningful from the project, two scientists from Sandia National Laboratories (SNL) were smiling ear to ear and immediately started asking me great questions.[52]

As a scientist, when the audience is asking interesting and thoughtful questions, you know that the talk went well.

After we ran out of time, I was asked to join Dr. Steven Conrad and Paul Kaplan from SNL for lunch where we discussed numerous aspects of national security in depth and ways that we could use science to improve national security and policy.

As I was nearing the end of the completion of my PhD coursework at UMN, I was entering the "all but dissertation phase" of the program, had passed my qualification exam, and was beginning to look for work as the bulk of my thesis and research had already been published in peer-reviewed journals.

During this time, I designed a blockchain for supply chain verification and validation, which few people understood at the time, and I was also working on building a software platform to map the links between critical infrastructures. I then took the show on the road and presented these new ideas and technologies to the leadership of the biggest companies in food and agriculture to obtain buy-in from these stakeholders.

The biggest challenge when deploying any new technology or platform aimed at large companies and enterprises is obtaining their support and agreement to use the product. Through these meetings, I received numerous offers for employment post-graduation, and I knew that I had to select soon.

Oddly, my acting center director, Dr. Amy Kircher, was suggesting and pressuring me to work for the State Department, which I thought was strange place for a scientist in national security to work, and one which I felt would result in a dead-end career path.

How little I understood at the time.

The conversations with Dr. Steve Conrad and Paul Kaplan continued, and they received permission to send me sensitive but unclassified information related to complex systems in national security, which further piqued my interest in working at SNL.

Eventually, I was asked to apply to SNL, and on-site interviews in Albuquerque, NM were quickly arranged.

A few weeks later, I flew down to SNL which is located on Kirtland Air Force Base. Since Sandia's primary mission is the construction of nuclear weapons, the facility is one of the most tightly guarded US government complexes in the world.

To enter the base, I had to be first granted permission and cleared outside the base, then I entered the base where everyone is subject to a security inspection, and, once inside the base, I always had to be escorted. I found all of this to be tremendously exciting.

The interview process was grueling and, to this day, rivals my most challenging interviews. I had group and individual interviews from 8 a.m. to 5 p.m., and I was required to give a national security focused technical presentation halfway through the day. The interview questions were highly technical and ranged from signal processing on electromagnetic signals to identify a person's location indoors, to determining how to structure and analyze financial transaction data to identify operational nodes in a network (I later learned that these were tools and methods developed by the interviewees).

At the end of the day, I was mentally exhausted, and the feedback from Paul and Steve indicated that I had done well.

About three weeks later, I received an offer from SNL, which I promptly accepted, and I was to begin work a few months later.

Back in Minnesota, I scheduled a meeting with Dr. Bender and told him that I was ready to defend my dissertation and that I was accepting a position at SNL and would begin work there in a few months. I think he was surprised and happy for me at the same time.

I think he was surprised because PhD students tend to linger around the university and not complete their degrees quickly, and I had completed the program at a lightning pace. The only difficulty remaining was attempting to find an available time to schedule all my thesis committee members, which ended up being several months later. The saddest two moments of my PhD occurred as my departure from NCFPD was eminent.

The first was that Chinese government officials were visiting the University of Minnesota and our research center to discuss collaborating on mitigating a long list of biological and chemical threats to the food and agriculture industry.

In recent history, the Chinese had used melamine to fake protein tests in baby formula and pet food, leading to the deaths of thousands of infants and millions of pets in the United States, including my own Siamese cat named Tessa.[53] [54]

Also, the Chinese were suspected of importing Porcine Epidemic Virus into the pork production system in the United States, potentially as a form of economic warfare, among many other suspicious, nefarious, or unscrupulous activities which harmed Americans.

I refused to participate in any of the meetings with the Chinese as I didn't see the point in meeting with potential adversaries that seemed to not have any interest in acting in good faith. At the end of their visit, I was forced to take a group picture with Col. John Hoffman and the Chinese. From the expressions on our faces, neither one of us was happy to be in the picture.

I hope someday that our relationship with the Chinese improves, but I am skeptical based on their government's behavior.

The move to Albuquerque went amazingly well. I found and bought a small house in a nice neighborhood in the northwest part of town with easy access to the hiking trails in the beautiful Sandia Mountains within a week of moving to New Mexico. SNL paid for contractors to pack, move, and unpack my items, which was the first time I ever experienced this luxury.

My first days at Sandia were typical of corporate America, and I spent these days filling out paperwork, registering for benefits, and sitting through endless pre-recorded presentations on safety, security, workplace fairness, and workplace harassment. When these tasks were finally completed, I was assigned to various projects throughout my organizational unit, which was mainly focused on modeling and simulating high consequence events or disasters related to national security.

It was my experience that each scientist or engineer in my organization at SNL had quite a bit of flexibility in the tasks on which they chose to work. This was not the case for engineers working on atom bombs; they worked on the same tasks, day in and day out.

The other unique aspect of my organization at SNL was that we had the ancillary task to respond to any natural disaster or security threat in the United States. If there was an attack against our country, and the government required emergency scientific and engineering expertise, then we had the facilities to operate continuously. With food, beds, and showers available, we could save valuable time, while being protected in the military installation, to address any national security problem we were facing.

Initially, I was assigned to a desk in a non-classified space where I shared an office with another rock star engineer. We were both waiting for our Department of Energy "Q" clearances to be granted. A Department of Energy (DoE) Q clearance is equivalent to a Department of Defense Top Secret clearance. The modern-day clearance system was created during World War II and was finalized with the Atomic Energy Act of 1946.

The original intent of the classification was to protect and keep secret advanced weapon research projects like the Manhattan Project atom bomb. During World War II, many of the scientists and engineers developing the atom bomb were concerned that a person in the government or military could go rogue and use an atom bomb nefariously. Some scientists were also concerned that if personnel outside the research and development teams could have access to atomic weapon systems information that this information could fall into enemy hands, and these concerns were echoed by personnel in the federal government.

The result was that DoE Q clearance holders were given access to restricted data (information pertaining to nuclear weapons design and development under special access programs) and DoD Top Secret clearance holders do not have access to restricted data. This small caveat might not seem like a big deal to most people in the security clearance world, and most people holding security clearances are probably not aware of the differences, but the potential fallout from a DoE Q clearance holder leaking information is "exceptionally grave" or "inestimable." For these reasons, I will not go into any specific detail about what I did or learned while employed at SNL, only that my experience at SNL gave me great insight into the many threats, vulnerabilities, and consequences facing our nation.

While waiting for my Q clearance, I was assigned to a wide variety of sensitive but unclassified projects for the usual mix of three letter agencies involved in security issues. I found this type of work to be largely boring as much of it pertained to mitigating natural disasters. At SNL, any work in public health, biological domains, or weapons—and thus of interest to me—was typically classified. I spent a few hours per day meeting with people working across the laboratory in various capacities related to bioweapons, biowarfare, and biosecurity.

In these meetings, I met one of the founders and creators of the BioWatch program.[55] The BioWatch program is a program where several US government agencies use a variety of methods to conduct biosurveillance within the United States; namely in large cities, in mass transportation systems, and at crowded mass gatherings. The general idea is that BioWatch collects biological samples from the environment with people walking in crowds with specialized equipment, from stationary devices placed in areas concentrated with people like subway platforms, or, in some cases, from aircraft with specialized spectral analyzing equipment that can scan the air for biological or chemical contaminants.

The main problem with BioWatch is that it relies on PCR testing, which takes time to run and analyze the results. By the time an attacker has spread the agent to high enough concentrations for a detection method to capture a sample containing the agent, and then have the agent tested in a laboratory, it is too late. People would already be exposed, shortly be infected, and potentially spreading the disease to others.

This is all dependent on which agent was used by the attacker.

This is the fundamental problem of many detection methods used to detect bioterrorism. The difference in time between people showing up at the emergency room presenting with illnesses and the time that the detectors and labs identify an attack does not give the defender a tactical advantage. The biosurveillance system may only provide two to three days' advanced warning in optimal conditions, and probably fewer days or hours in practice.

While these types of meetings were expanding on the knowledge that I had gained during the previous four years of my education and experience, I found it helpful to meet these people and learn from their experience and the challenges they faced when designing and implementing these programs.

As a PhD student, before I worked at Sandia, I began assembling a list of every known biological and chemical agent on the planet to find gaps in what could be obtained easily and used effectively as a weapon to kill the largest amount of people possible without being detected. The idea is that if you know what can be used as an effective weapon, then you can prioritize resources to mitigate these threats.

The fact is that local, state, and the federal government are resource-constrained, and there is an infinite list of potential threats to mitigate against. If you can identify what the highest risk agents to be used in a biological or chemical attack are, then you can hopefully develop simple and cost-effective methods to mitigate the risks.

While this makes sense to most people, I quickly learned that, when it comes to national security, the politicians or bureaucrats in Washington, DC, do not care about reason or logic.

* * *

After 9/11, the US government implemented a program via the DHS to provide grants to state and local governments to prepare and mitigate national security threats. As some of you might recall, many local governments were gifted with armored vehicles from the DoD via DHS, and these types of programs were just the tip of the iceberg.

DHS conducted a bi-annual data call, where state government agencies were to submit data related to all the critical infrastructures in their states that were critical to their state and provide justifications to the federal government as to why these critical infrastructures should be protected.

After meeting with numerous state and local government employees, it was clear that they viewed the data call to obtain federal grant dollars, and some of these people were excellent creative writers and some were also able to get industry to support their needs assessments.

Once DHS and the related supporting agencies completed their independent review, with the aid of government contractors, large grants were issued to state governments with no strings attached. Some states mitigated security risks, and others bought excess firetrucks, police cars, and other equipment and stored them in garages until the equipment that they were using met the end of its lifecycle.

Could you imagine that a significant agroterrorism risk was identified in your state, and the federal government gave your state funds to mitigate that risk, but then the state simply dumped the cash into the general fund for normal spending?

This is what was happening then and is still likely happening to this day.

The more of these types of problems I witnessed in the security industry, the more jaded I became with the scientific process and science's role in national security. I was learning quickly that government scientists and academics at universities working in security were all fighting over peanuts to mitigate threats and solve the problems which some elected idiot or political appointee determined to be important.

For example, I was sitting in my mentor Paul's office at work when we received a call from a US congressman related to a potential national security threat the congressman had apparently imagined over the weekend. The congressman was gravely concerned that this threat was likely to happen and wanted us to get to work on the problem immediately.

That is when Paul said something to the congressman that I will never forget: "Science cannot solve your policy problem."

The phone went silent, and the congressman hung up. Paul also hung up, laughed, and said that he had been using that statement for over twenty years when dealing with politicians and bureaucrats. He was correct: science—the process of hypothesizing, collecting data, testing, analyzing, and interpreting results—does not have anything to do with policy.

Often, policy makers and bureaucrats think that science and engineering are some magic potions that can make all the problems in society go away.

The opposite is typically true; when scientists and engineers embark on a problem, we often discover more problems or break our prototypes, thus requiring more time and resources to investigate and solve the new problems, weak prototypes, or fill newly discovered data gaps.

This creates more problems for the funders of research, drives up costs, and increases project timelines.

Once I was finally cleared (after about five months), I began receiving threat intelligence for a variety of topics including biological threat and public health intelligence. In my opinion, none of the information that I was briefed warranted any type of classification. Some of the classified information that I was briefed on I already knew and learned during graduate school.

Most of the classified threat intelligence that I was exposed to could be easily obtained by sitting at a computer connected to the internet, asking the right questions, and compiling the information oneself. This is the stupidity of "intelligence" classification.

The worst part of having a clearance is that once you are told that something is classified you are barred from talking about it, even if that classified knowledge is leaked by someone else and becomes public knowledge. The problem of over-classification in the US government is a significant problem to a free society.

If the public cannot be informed as to the threats facing them, then how can they or their elected representatives take the necessary actions to decide the best course of action to mitigate threats or to protect themselves or their families? If the public is uninformed, then we cannot solely rely on the government to protect us from threats.

The other thing that happened once my clearance was finalized was that I was briefed on several Q level Operations and Support Special Access Programs (SAPs).

SAPs are established for a specific class of classified information that imposes safeguarding and access requirements exceeding those normally required for information at the same level.

I was fortunate enough to not be assigned to work directly in these programs and nor did I want to work in them. As a clearance holder, more access to more programs comes with more personal scrutiny into your personal life and all your behaviors.

Who really wants that?

Knowing about these programs, their objectives, and how they worked to a degree was more than enough information to understand what was going on in these programs. Additionally, a person can review the government's contract solicitations from the entities working on the SAPs to understand the goals and intent of the programs.

I mainly used the information I learned to be strategic in how I worded my proposals related to pandemic preparedness and response, biowarfare, bioterrorism, biosurveillance intelligence collection, and how to communicate and pitch my ideas to potential corporate and government project sponsors.

If you know the same secrets that your potential sponsors know, and you know what they really want but can't say, then you are likely to have greater success in satisfying their needs and achieving your goals.

Namely, obtaining the funding.

Despite the classified aspect of my duties at the lab, I worked on pandemic preparedness and response planning, provided expert technical assistance to the leadership of foreign governments, examined the relationships of supply chains and their relationship to emerging infectious diseases in complex supply chains, prevented the spread of infectious diseases (biosecurity and biosafety), optimized patient movement in hospitals to minimize infectious disease transmission, identified and analyzed points of failure in complex and interconnected critical infrastructure systems, predicted the impacts of various disasters or attacks on healthcare and food systems, worked with the government and corporations to prevent using the food system as a WMD delivery vehicle to kill hundreds of thousands of people in a single attack, and provided classified and non-classified briefings to project sponsors and stakeholders in the US government when called upon or when I felt they were necessary to convince stakeholders that there were newly discovered security threats or vulnerabilities that required mitigation.

Over time, I became quite proficient in using super computers to build models and using simulations to optimize the national pandemic preparedness and response plans, which were primarily based around highly transmissible pandemic influenza scenarios and were communicated to the highest levels of US government agencies involved in pandemic preparedness and response.

Most of SNL and the numerous other agencies and laboratories working on the issues, findings and recommendations must have been tossed in the trash around October 2019.

As I had been rapidly learning how the military medical complex worked, I was increasingly becoming disheartened with the state of national security and how little progress was being made externally and internally to my areas of expertise.

As a joke, several of my friends who are experts in security and I routinely said that we have made backwards progress in national security since 9/11, and it has cost US taxpayers billions of dollars.

In 2013, the funding for impactful national security work was drying up due to the austerity measures enacted by Congress (a.k.a. sequestration).

The government was shut down, and while the premier nuclear weapons laboratory was high up on the food chain in terms of government spending priority and had acquired a large operational nest egg to keep the lab operational during funding squeezes, most of us at the lab would have been eventually laid off if the government shutdown would have lasted only a few more weeks.

For any civil servant, government shutdowns can be highly stressful. This caused several excellent scientists and engineers at SNL to begin looking for work elsewhere and many were quickly gobbled-up by the tech giants in Silicon Valley over the next few months.

The morale among laboratory staff members was somewhat morose.

The final straw for me was that my own independent work was increasingly at risk of classification without good reason.

Unfortunately for me, the manager that I was assigned to really didn't understand anything about science, engineering, or security, but was somehow qualified to supervise PhD level staff. This person was also my derivative classifier.

A derivative classifier is the person that reviews information to determine whether it should be classified and at what level it should be classified. Suddenly, all my work was coming under increased scrutiny to be classified and I had to argue with a person that was not qualified for their position as to why each piece of my work should or should not be classified.

A perfect example was the work I did which examined the effect that a pandemic would have on the food supply.

From my experience working in supply chains and interconnected systems, I observed that many US critical infrastructures were fragile due to their reliance on just-in-time economy profitability and many single points of failure, which are necessary to sustain the functioning of these systems.

At SNL, we frequently examined these critical interdependencies under various scenarios, and I took it upon myself to investigate the effects of worker absenteeism on these critical infrastructures.

When I submitted a draft of the peer-reviewed manuscript for approval to be released outside the lab, my manager called me into a series of meetings where I had to argue that this information should not be classified because pandemics naturally occur, and that society should be aware of the consequences of a fragile food system because, in a severe pandemic, many of us would starve to death.

Thankfully, I won this battle, and the manuscript was approved for release.

I was not so effective in other arguments related to classification, and I could read the writing on the wall. If I were to stay a scientist at SNL, then my work would be increasingly classified and underfunded, and I would be stuck in the government for the remainder of my career.

So, I decided to begin looking for work elsewhere in late summer 2014.

***

The first week into my job search I discovered an opening at an organization called EcoHealth Alliance (EHA) which was based out of New York City.

The position was specifically looking for an infectious disease scientist or epidemiologist with a technical and quantitative background to lead several projects related to biosurveillance and digital disease detection technology with the job title of Senior Scientist.

While reading about the organization, I fell in love with the mission.

At the time, the mission was to prevent emerging infectious diseases and pandemics by focusing on conservation.

The idea, which had face validity, was that if you can protect and conserve wildlife and the environment, then you can prevent pandemics by reducing the exposure rate of humans and domestic animals to wildlife which carry zoonotic diseases. Just from reading the website, I felt like the people that worked and ran EHA really understood disease transmission dynamics and, since I am a conservationist at heart and spend much of my free time in nature, I felt that the organization aligned with my values.

I updated my resume and submitted it with the other required application materials via email directly to the president of EcoHealth Alliance, Dr. Peter Daszak, the man I would later realize was directly responsible for our greatest modern pandemic.

CHAPTER 8

Leaving the Military Industrial Complex, or So I Thought...

Only a few days had passed from the time I submitted my application to EcoHealth Alliance until the time I was contacted by a man named Dr. Aleksei Chmura to schedule a phone interview with Dr. Peter Daszak. Aleksei was the chief of staff at EcoHealth Alliance.

During the interview, Peter was kind and charming, and his British accent gave him an air of sophistication. He was passionate about EHA's mission to prevent pandemics via conservation, and I loved everything that he told me about the organization's mission and the role. Peter asked me about my research interests and experience, all typical of a first-time call with a hiring manager.

We ended the call, and it was my impression that the call went very well as we bonded over our love for nature and wildlife.

A few days later, I was driving home from purchasing groceries when I received a call from Peter with Aleksei on the line where they invited me to come to EcoHealth Alliance's headquarters in New York City the following week. They offered to arrange my airfare and lodging, and I made the necessary arrangements to travel to the Big Apple.

I had previously visited New York City (NYC) a few times for work and was thus somewhat familiar with Manhattan and Brooklyn, and was excited to interview and eat some great food, even if the position ended up not being a good fit.

My longtime girlfriend Emily decided to travel with me to NYC as she had numerous family and friends in town and could catch up with them while I was focused on my interview. We flew in the day before my interview at EHA, and I refreshed myself on the specifics of emerging infectious disease ecology, ecological niches for infectious diseases, and epidemiological modeling of infectious disease.

The next morning was a gorgeous, warm, sunny, fall day. I decided to leave early and walk from our hotel on the lower end of the Upper East Side to EcoHealth's office near Hudson Yards, which was under heavy construction at the time. I arrived about thirty minutes too early, so I went into the church next to EcoHealth and sat in the pews and read my Control of Communicable Diseases reference manual until my scheduled interview time.

Five minutes before my interview, I walked into EcoHealth's building on West 34th Street and took the elevator to the seventeenth floor. As I walked in, I thought that the office was a little dingy for a company located in downtown New York. The office smelled slightly dank, had puke colored carpet, and the walls were off-white with yellow accents.

The senior staff offices were like fishbowls, and I was shown my potential new office, which had windows and a view of the neighboring buildings where I was told that falcons perched on the air conditioners and would feast on pigeons.

The interview structure was typical of academic positions; I was to meet with my future subordinates in the department, interview with other scientists, meet with the hiring manager (Peter), and give a scientific presentation around the lunch hour which anyone from the organization could attend.

The conversations were fun and interesting, and I could see how all the scientists and leadership worked together as a team.

Scientific presentations for job interviews are always challenging, as sometimes you are not aware of the of the potential political landmines related to your candidacy, so it's best to give a presentation where you touch on a variety of topics so that every person in the audience likes or relates to at least one of your past areas of research, methodology, or perspectives.

I presented on my past work including infectious diseases in food systems, optimization of pandemic preparedness and response at the national level (vaccination, social distancing, strategy, drug treatments, model parameterization, policy), and spatial modeling of emerging infectious disease spread in geographic information systems. I received a few difficult questions, which I handled well, and everyone in the room looked quite impressed.

After my presentation, I had a brief meeting with Peter where we thanked each other, and then I departed EHA for my hotel. As I walked out of EHA, my overall impression was that the organization was growing and that my technology and infectious disease skills were a good fit for the role.

Not only was I worried about my future if I continued to work in the military industrial complex as a scientist, but Emily and I were quickly growing tired of the island that was Albuquerque. Albuquerque was a dangerous and violent city and felt like a bigoted and narrow-minded town of roughly 500,000 people.

The New Mexican and Mexican food (I learned the differences between the two while living there) were both incredible but finding an edible fillet of fish at the grocery store was impossible. There was a serious lack of culture and education in the state, and most of the people with disposable income were in some way affiliated with the laboratory, the government, or the military.

This made it almost impossible to get away from work no matter where you went.

Often, the first question anyone asked is where you worked, and if you said the laboratory, they automatically assumed that you were an engineer.

Emily and I were ready to inject some culture back into our lives.

A few days after my interview, I received a call from Peter, and he offered me the job in New York. He was adamant about having me start the job as soon as reasonably possible. I explained to him that I owned a house in Albuquerque and that moving to New York would take some time.

We quickly negotiated salary, relocation benefits including a house hunting trip, and an annual nondiscretionary budget allocation to be used at my disposal. Peter offered to begin paying me immediately, which I later learned was likely the first sketchy financial transaction that occurred during my time at EcoHealth Alliance.

Nonetheless, I signed the job offer on September 23, 2014.

I valued and appreciated Peter's ability to act quickly and decisively, as that is a rare quality in scientific leaders globally.

Most scientific leaders spend far too much time pontificating and prevaricating—a disastrous quality and behavior when possessed by any leader.

The next week Emily and I traveled to New York City where we had only a few days to find an apartment in an overheated real-estate market.

Twenty and thirty-somethings were flocking to the city in droves and properties for purchase or rent were only on the market for a few days before they
would be rented or sold.

We looked at residences all over Manhattan, and we finally settled
on a swanky luxury building on 45th Street in Hell's Kitchen, which was
about a fifteen-minute walk down the West Side to my office at EHA. For
Manhattan, we had many comforts in our new building, laundry in our
unit, a "large" modern kitchen, and sixteen-foot-high ceilings with a view
overlooking the Hudson River.

The combination of food, nightlife, and culture was a much-needed
change in our lives.

Moving to New York City was no easy task and EHA was not interested
in managing the move, so Peter cut me a large check to handle the move on
my own. We were going from a moderate sized three thousand square foot
home to an eight hundred square foot condo. We decided to sell what we
could in New Mexico and move all our belongings out to the East Coast.
This included a library full of books, which was a mistake I still deeply
regret to this day.

For the move, I purchased a large eighteen-foot enclosed trailer, and
I pulled the trailer with my Jeep from New Mexico right into downtown
Manhattan.

Thank God for my life experience with towing heavy equipment,
including my military experience.

Upon arrival we hired movers to unpack our trailer and move our items
into our unit. We ended up buying new furniture in NYC and putting most
of our belongings into storage outside the city to save money.

Being in New York City during its most vibrant time in modern history was exhilarating, and I thrived in the environment of specialization.
During this period, the population of people in their twenties and thirties
had exploded, and the city was peaking in economic, commercial, and residential growth. There was no need to cook, clean, or perform any other task
other than focusing on work.

The nature of the extensive service economy in the city made it possible
to be highly productive, creative, and focused. I quickly made new friends
from all walks of life that worked in music, theater, television, fashion, telecommunications, event production, high food service, law, consulting, and
finance.

When work ended, there was always something exciting to do or a new restaurant to dine at, which was an invigorating driving force to work very hard. Many people I knew worked hard and played harder.

CHAPTER 9

Understanding the Risk of Working at EcoHealth

My first few days at EcoHealth Alliance were not like other jobs that I had in the past.

There were no equal opportunity presentations to sit through, or videos on laboratory environmental health and occupational safety to click through.

There really was not any type of onboarding process at the company. On my first day, I had a meeting with Peter where he candidly told me what my primary objective was: correct a failing department within the company and obtain additional funding from the Department of Defense.

He told me that the previous scientist and leader, Dr. Niko Preston, left on bad terms.

I took this information with a grain of salt and the knowledge that some of the employees in the department were disruptive to the team and the organization. Under Dr. Preston's leadership, the department had received a large multi-million-dollar contract from the Defense Threat Reduction Agency (DTRA),[56] which is a sub agency of the DoD.

I found it hard to believe that Dr. Preston was underperforming in some way.

Over time, I observed that Peter routinely trashed anyone that did not agree with him or with his objectives, which is often a trait of people in leadership positions that are unqualified to lead and are insecure about their position as leaders. As a leader there is nothing to be gained by admonishing a former employee to a new employee and can harm your role as a leader by undermining your ability to provide purpose, direction, and motivation to your subordinates to accomplish the mission.

DTRA's mission is the protection of the warfighter, and I had previously worked with the agency while employed at SNL. I was also familiar with DTRA from my time in the military. DTRA is known as a "force multiplier" which enables the DoD, the US government, and the US government's international partners to counter and deter Weapons of Mass Destruction (WMD) and emerging threats (i.e., chemical, biological, radiological, and nuclear threats).

The purpose of the contract that my predecessor had obtained was to build digital disease detection tools. The concept of digital disease detection tools was that by analyzing digital information signals, we could identify infectious disease threats and outbreaks before they were detected via more traditional laboratory based infectious disease surveillance methods, like collecting blood samples to test for disease antibodies or testing the blood for the presence of the disease itself (i.e., polymerase chain reaction tests, a.k.a. PCR).[57]

A major problem in public health at that time, which still exists to this day, is that by the time a highly transmissible infectious disease has sufficiently spread to the point where it can be noticed or detected as being unusual, it is incredibly difficult to mitigate, prevent, or stop further spread of the disease (a.k.a. disease transmission).

For every case that you detect with a biosurveillance system, there are likely many more, if not hundreds or thousands of, undetected cases circulating in the population. This is one of the many weaknesses of laboratory-based infectious disease surveillance systems. This weakness of laboratory-based surveillance systems exists for several reasons, and one of the most important factors is caused by the phenomena that some people with strong immune systems and in good physical condition, when infected by an infectious disease, only have mild symptoms, or subclinical illness (i.e., no easily detectable symptoms), and can shed the disease into the environment and transmit the disease to others.

Also, many people that become mildly or moderately ill never go to the doctor or hospital where their illness would be diagnosed by a medical doctor. Their diagnosis could then be aided by the analysis of a biological sample, which could be tested for the presence or indicators of an infectious disease.

Often, medical doctors only perform a differential diagnosis of the patient's illness and diagnostic tests are not ordered by the medical doctor. I am sure many of you can relate to this experience first-hand.

This is known as bias and, in epidemiology jargon, as *surveillance bias*.

This medical process of diagnosis, without testing, is another inherent weakness in laboratory-based infectious disease surveillance systems and you cannot detect or respond to an infectious disease outbreak unless you have information that it is occurring. Occasionally, astute medical doctors and clinicians notice a strange rapid increase in the incidence of infectious disease cases presenting at hospitals and notify public health authorities, but this is an exception and not the norm.

Lastly, even if the public health and medical system collects and tests enough biological samples, there must be a sufficient number of cases of the disease to trigger an alarm, and how sensitive that alarm is depends on 1) the natural occurrence or prevalence of each specific disease; 2) the time frame in which the disease occurs; 3) the number of samples collected and tested; 4) how the samples are selected and sampled from the population (i.e., random or non-random); 5) the accuracy and reliability of the diagnostic tests being used; and 6) the geographic concentration of the positive cases.

The goal of the DTRA contract that I was assuming ownership of was to develop advanced signals intelligence technologies and systems that could analyze open-source digital data across all information streams to detect infectious diseases globally in near-real time.

As I took over the contract and began to manage the project, I noticed something strange. My first two months of salary, when I was not working on the contract, were billed against the DTRA contract that I assumed. Since I was previously a contracts technical representative for the US government, I knew that this was not allowed under the Code of Federal Regulations (CFR). I questioned why Peter and EcoHealth would do something so obviously incorrect?

I would later find the answer to this question myself.

The department that I was assuming control of, Data and Technology, consisted of mostly remote employees that had skills in big data processing, machine learning (a.k.a. artificial intelligence), and software engineering. One of the major failings of the department that I quickly identified was that there was a fundamental lack of knowledge about how health care, public health, and infectious disease surveillance systems worked in the real world.

Also, the employees seemed to have no understanding of what public health practitioners did and how they used infectious disease surveillance systems to detect and respond to infectious disease outbreaks.

Luckily, I had the right skills and experience to correct these deficiencies as I quickly realized that I did not ask the best questions to identify these problems during the interview process.

Another problem that I identified was that Dr. Daszak had been using the Data and Technology team to work on all his hair-brained ideas and side projects. Instead of working on the DTRA contract and making progress toward the Statement of Work (SoW), he would ask the employees to begin work on highly involved and complicated software development and data engineering projects, while billing the employees' salaries to the government.

I believe that he engaged in this behavior because Peter lacked consideration for other's time and the only objective that was important for the entire company, at any moment in time, was whatever popped into Peter's brilliant mind at a moment's notice.

For example, Peter would curl his arms up like a tyrannosaurus rex and make a typing-on-a-keyboard gesture while requesting that my department organize, analyze, and build back-end and front-end software for a hundred years' worth of infectious disease emergence data, while saying something like "You think you can have that done in a week?"

This demeaning behavior was generally representative of Peter as a leader and was a behavior that I had previously identified in others whom I classified as assholes, idiots, or both. He simply didn't understand technical or quantitative research and development, and I believe these behaviors were representative of an inferiority complex he had with anyone that had greater technical or quantitative abilities.

The most significant problem with Peter siphoning my employees to work on his valueless side projects was that he was committing fraud against the government by billing their time to DTRA. Despite Peter's inability to lead and conduct business legally, I had to find a way to communicate and persuade Peter that the DTRA contract was the top priority, which was an easy argument to make.

In the absence of effective leadership, the staff that I had inherited seemed to work on whatever they felt was important instead of working to make our customer, DTRA, happy and excited about our work. My department's personnel that were objecting to doing the work outlined in the contract and SOW were vocal in their objections.

I quickly identified the problem children that Peter had previously alluded to in my first meeting with him and made plans to terminate them or force them out of the company.

During the first few months of a new job, it takes some time to understand the entirety of the business and social system that you have been inserted into.

Since I was the only card-carrying PhD in public health at EcoHealth Alliance, and since I had already been successful in obtaining government funding and grants from several agencies and private companies, I was asked to consult and review several grant and contract proposals during their planning, design, and implementation phases. In academia and professional research and development firms, this is a normal and typical business process.

The more eyes and brains that review and criticize a proposal, the greater the chances of success on receiving an award and funding. The more people that you include as co-investigators in a proposal, the more the proposal is typically valued by the project's sponsor or government agency for a variety of reasons, which are often political in the sense that the government likes to get the most bang for their buck and that they like to see that the money is being spread across special interests and congressional districts.

One of the first proposals that I reviewed at EcoHealth was a proposal that was in preparation for submission to the National Institutes of Health with a long list of collaborators, as it was explained to me by Dr. Daszak. I later found out that the proposal was likely being edited or renegotiated after submission when it was provided to me for review. The listed investigators were Peter Daszak from EHA, Zhengli Shi of the Wuhan Institute of Virology (WIV), ShuYi Zhang of the East China Normal University, Changwen Ke of the Chinese Centers for Disease Control and Prevention of Guandong Province, Jonathan Epstein from EHA, Kevin Olival from EHA, XingYe Ge of the WIV, Guanjin Zhu from the Guangdong Entomological Institute, and Yun-Zhi Zhang from the Yunnan Center for Disease Control.

The proposal received a glowing Letter of Support from Dr. Ralph Baric, a microbiologist and virologist at the University of North Carolina (UNC).

The NIH grant proposal was titled: "Understanding the Risk of Bat Coronavirus Emergence."[58]

Research proposals submitted to NIH are scientifically dense and are often page limited. So, the person or group of people submitting the proposals use wide margins, small font, and attempt to cram as much information as possible into the grant proposal.

As COVID gain of function history demonstrates, the benefit of the NIH RoI proposal structure and review process to obtaining government funding does not result in a good value to the taxpayer. In my opinion, the NIH proposal and review process should be completely overhauled and replaced with something like the DARPA grant proposal and submission process. DARPA's submission process requires the investigator to focus on the potential return on investment, the likelihood that the goals will be achieved, the risks of the project, the length of time required to achieve the project's goals, and the difference or impact the investment will make compared to the status quo (this is known as Heilmeier's Catechism).[59]

Nonetheless, this is an exact copy of a section of the grant submitted to the NIH by both EHA and the WIV titled "Specific Aims:"

SPECIFIC AIMS

Zoonotic coronaviruses are a significant threat to global health, as demonstrated with the emergence of severe acute respiratory syndrome coronavirus (SARS-CoV) in 2002, and the recent emergence [of] Middle East Respiratory Syndrome (MERS-CoV). The wildlife reservoirs of SARS-CoV were identified by our group as bat species, and since then hundreds of novel bat-CoVs have been discovered (including >260 by our group).

These, and other wildlife species, are hunted, traded, butchered and consumed across Asia, creating a largescale human-wildlife interface, and high risk of future emergence of novel CoVs. To understand the risk of zoonotic CoV emergence, we propose to examine 1) the transmission dynamics of bat-CoVs across the human-wildlife interface, and 2) how this process is affected by CoV evolutionary potential, and how it might force CoV evolution.

We will assess the nature and frequency of contact among animals and people in two critical human-animal interfaces: live animal markets in China and people who are highly exposed to bats in rural China. In the markets we hypothesize that viral emergence may be accelerated by heightened mixing of host species leading to viral evolution, and high potential for contact with humans.

In this study, we propose three specific aims and will screen free ranging and captive bats in China for known and novel coronaviruses; screen people who have high occupational exposure to bats and other wildlife; and examine the genetics and receptor binding properties of novel bat-CoVs we have already identified and those we will discover. We will then use ecological and evolutionary analyses and predictive mathematical models to examine the risk of future bat-CoV spillover to humans. This work will follow 3 specific aims:

Specific Aim 1: Assessment of CoV spillover potential at high risk [*sic*] human-wildlife interfaces. We will examine if: 1) wildlife markets in China provide enhanced capacity for bat-CoVs to infect other hosts, either via evolutionary adaptation or recombination; 2) the import of animals from throughout Southeast Asia introduces a higher genetic diversity of mammalian CoVs in market systems compared to within intact ecosystems of China and Southeast Asia; We will interview people about the nature and frequency of contact with bats and other wildlife; collect blood samples from people highly exposed to wildlife; and collect a full range of clinical samples from bats and other mammals in the wild and in wetmarkets; and screen these for CoVs using serological and molecular assays.

Specific Aim 2: Receptor evolution, host range and predictive modeling of bat-CoV emergence risk. We propose two competing hypotheses: 1) CoV host-range in bats and other mammals is limited by the phylogenetic relatedness of bats and evolutionary conservation of CoV receptors; 2) CoV host-range is limited by geographic and ecological opportunity for contact between species so that the wildlife trade disrupts the 'natural' co-phylogeny, facilitates spill-over and promotes viral evolution. We will develop CoV phylogenies from sequence data collected previously by our group, and in the proposed study, as well as from Genbank. We will examine co-evolutionary congruence of bat-CoVs and their hosts using both functional (receptor) and neutral genes. We will predict host-range in unsampled species using a generalizable model of host and viral ecological and phylogenetic traits to explain patterns of viral sharing between species. We will test for positive selection in market vs. wild-sampled viruses, and use data to parameterize mathematical models that predict CoV evolutionary and transmission dynamics. We will then examine scenarios of how CoVs with different transmissibility would likely emerge in wildlife markets.

Specific Aim 3: Testing predictions of CoV inter-species transmission. We will test our models of host range (i.e. [*sic*] emergence potential) experimentally using reverse genetics, pseudovirus and receptor binding assays, and virus infection experiments in cell culture and humanized mice. With bat-CoVs that we've isolated or sequenced, and using live virus or pseudovirus infection in cells of different origin or expressing different receptor molecules,

we will assess potential for each isolated virus and those with receptor binding site sequence, to spill over. We will do this by sequencing the spike (or
other receptor binding/fusion) protein genes from all our bat-CoVs, creating
mutants to identify how significantly each would need to evolve to use ACE2,
CD26/DPP4 (MERS-CoV receptor) or other potential CoV receptors. We
will then use receptor-mutant pseudovirus binding assays, in vitro studies
in bat, primate, human and other species' cell lines, and with humanized
mice where particularly interesting viruses are identified phylogenetically, or
isolated. These tests will provide public health-relevant data, and also [sic]
iteratively improve our predictive model to better target bat species and CoVs
during our field studies to obtain bat-CoV strains of the greatest interest for
understanding the mechanisms of cross-species transmission.

For most of the people reading this book, I suspect much of the information
detailed in the Specific Aims of the proposal submitted to NIH is beyond
easy comprehension.

That is by no means meant to be an insult.

Some of the work being done in infectious disease research called *gain
of function* is highly complex and nuanced. Gain of function research is pur-
posefully enhancing the pathogenicity, infectivity, virulence, survivability,
or transmissibility of an infectious agent.

Simply, that means making an infectious agent more dangerous.

There are numerous ways that this can be done in a laboratory, and the
easiest and simplest way to execute gain of function work is to merely collect
numerous samples from mammals thought to be a host for an infectious
agent (e.g., bats). Then, you select infectious agents (e.g., SARS-CoVs) from
the host that have the most dangerous traits (e.g., a bat may carry a single
virus or multiple viruses, each potentially carrying several viral phenotypes
in a single host or across multiple hosts).

Whether it be a viral or bacterial pathogen, all infectious agents
have some degree of genetic variability (i.e., genotypes). Some individual
microbes, virons, or other infectious agents might have slightly different
traits, which make them more dangerous, and once these dangerous traits
are identified in the sample, they can be selected and replicated. After the
most dangerous phenotype is identified and selected, it can be replicated in
a laboratory.

This process is known as cloning.

By cloning a dangerous virus in large quantity, humans artificially
increase the proportion of the selected phenotype compared to what occurs

naturally. Whether or not this type of gain of function work is safe or good for humanity depends on many details, and there are many additional sources of information related to the ethics of gain of function work.

This type of gain of function (GoF) work requires relatively little skill compared to what is described in the "Specific Aims of Understanding the Risk of Bat Coronavirus Emergence." This is an example of the simplest form and example of GoF work.

This work is being conducted all around the globe as you read this book right now, and this type of GoF work is the safest and poses the least amount of risk to life.

At the other end of the spectrum, there is the infectious disease gain of function work that is highly complex and complicated.

The GoF work managed and conducted by EcoHealth Alliance, the Wuhan Institute of Virology, and Dr. Ralph Baric at the University of North Carolina, which resulted in the creation of SARS-CoV-2, was the most sophisticated, complex, and complicated GoF research and engineering known to man.

The GoF work proposed by EcoHealth Alliance was first funded by the United States Agency for International Development (USAID), not by Dr. Anthony Fauci (NIH NIAID), like Senator Rand Paul and others have claimed.

In the proposal, "Understanding the Risk of Bat Coronavirus Emergence," there are details of how the project was initially funded and implemented. In the proposal, USAID is credited by both American and Chinese researchers listed in the proposal as providing the funding to form the relationship between American and Chinese scientists, and to collect the necessary biological samples from bats to obtain coronaviruses.

Without USAID's funding, the relationship between EHA, the WIV, and UNC would not have occurred, Dr. Baric's advanced methods and sophisticated biotechnology would not have been transferred to China, and, without USAID's funding, they would not have collected the first bat coronavirus samples in China.

Lastly, without USAID's funding, they would not have obtained the necessary preliminary data and would not have demonstrated that the collaboration between the EHA, the WIV, and UNC was an effective partnership to obtain NIH NIAID funding.

Without USAID's initial sponsorship of the GoF work, it is highly unlikely that SARS-CoV-2 would have been successfully engineered.

The research described by EHA does not explicitly say GoF, and that was intentional. On October 17, 2014, President Barack Obama's Administration declared a moratorium on gain of function research related to influenza and coronaviruses (Middle East Respiratory Syndrome [MERS] and Severe Acute Respiratory Syndrome [SARS]).

Here is exactly what President Obama's White House wrote:

Doing Diligence to Assess the Risks and Benefits of Life Sciences Gain of Function Research

Following recent biosafety incidents at Federal research facilities, the U.S. Government has taken a number of [*sic*] steps to promote and enhance the Nation's biosafety and biosecurity, including immediate and longer-term measures to review activities specifically related to the storage and handling of infectious agents.

As part of this review, the White House Office of Science and Technology Policy and Department of Health and Human Services today announced that the U.S. Government is launching a deliberative process to assess the potential risks and benefits associated with a subset of life sciences research known as "gain of function" studies. With an ultimate goal [*sic*] of better understanding disease pathways, gain of function studies aim to increase the ability of infectious agents to cause disease by enhancing its pathogenicity or by increasing its transmissibility.

Because the deliberative process launching today will aim to address key questions about the risks and benefits of gain of function studies, during the period of deliberation, the U.S. Government will institute a pause on funding for any new studies that include certain gain of function experiments involving influenza, SARS, and MERS viruses. Specifically, the funding pause will apply to gain of function research projects that may be reasonably anticipated to confer attributes to influenza, MERS, or SARS viruses such that the virus would have enhanced pathogenicity and/or transmissibility in mammals via the respiratory route.

During this pause, the U.S. Government will not fund any new projects involving these experiments and encourages those currently conducting this type of work—whether federally funded or not —to voluntarily pause their research while risks and benefits are being reassessed. The funding pause will not apply to the characterization or testing of naturally occurring influenza, MERS, and SARS viruses unless there is a reasonable expectation that these tests would increase transmissibility or pathogenicity.

The deliberative process will involve two distinct but complementary entities: the National Science Advisory Board for Biosecurity (NSABB) and the National Research Council (NRC) of the National Academies.

The NSABB will serve as the official Federal advisory body for providing advice on oversight of this area of dual-use research, in keeping with Federal rules and regulations. The NSABB will meet on October 22, 2014, to debate the issues and begin the process of developing recommendations.

Early-on in the deliberative process, the NRC will be asked to convene a scientific symposium focused on the issues associated with gain of function research. The NRC will also hold a second symposium later in the deliberative process, which will include a discussion of the NSABB's draft recommendations regarding gain of function research.

The NSABB, informed by discussion at the NRC public consultations, will provide recommendations to the heads of all federal entities that conduct, support, or have an interest in life sciences research. The final NSABB recommendations as well as the outcomes of the NRC conferences will inform the development and adoption of a new U.S. Government policy regarding gain of function research.

The broader life-sciences community will be encouraged to provide input through both the NRC and NSABB deliberative processes. The funding pause will end when the U.S. government has adopted a Federal policy regarding gain of function studies on the basis of the deliberative process described above, which is expected to occur 2015.[60]

Although this White House Press release from the Obama Administration was fairly written, the press release clearly states that there is a ban on gain of function research on SARS, MERS, and influenza viruses.

In EcoHealth's proposal to NIH in the Specific Aims section, Peter Daszak's best quality as a scientist spectacularly radiates: his writing ability. He has an impeccable understanding of rhetorical strategy, persuasive writing or argument, and consideration of his audience when developing prose. In the introductory paragraph of the Specific Aims section, the research proposed does not describe research that could be easily identified as SARS-CoVs gain of function, and that is an example of Peter's brilliance as a writer and strategist.

Before I explain the next three Specific Aims in lay terms, I will explain the process of how gain of function research is executed in slightly greater detail than previously described in this chapter.

The following description is intentionally simplified, and information is omitted to gradually increase your understanding of gain of function research and engineering.

By the end of this book, you should be able to understand exactly how the SARS-CoV-2 gain of function work was conducted.

Gain of function research has many steps. Each one of the steps described in the following paragraphs was necessary to successfully execute gain of function work at the time when this proposal was submitted (Fall 2016). Due to advances in synthetic biology, some of the steps used to conduct gain of function between 2016 and 2019 may no longer be necessary for gain of function work.

First, biological samples need to be collected from a species that are likely to be natural reservoirs for the target agent. In the case of this proposal, biological samples are blood, saliva, or feces, the reservoir is the bat, and the target agents are SARs coronaviruses.

To collect these biological samples, small teams are sent to the natural habitat site (i.e., research site) of the reservoirs (i.e., bats). Once the sample collection team arrives at the research site, they don personal protective equipment (e.g., bodysuits, goggles, respirators, gloves, and boots) to prevent exposure to infectious agents during sample collection.

Then, the research team enters the research site (i.e., caves with bats) and captures the bats with nets. As you might imagine, the bats are typically not too happy about this and will attempt to bite the researchers or will excrete urine or feces due to the stress.

Once captured, the research team will collect a small sample of blood with a needle and vacuum tube, obtain a saliva swab from the bat's mouth, and collect a sample of feces from the animal. All of this is labeled and indexed, and the collected samples are immediately refrigerated, if necessary.

Without the sample collection from bats prior to 2016, the gain of function work to build SARS-CoV-2 could not have occurred, as they would not have had the necessary genetic material to engineer the agent. In this case, sample collection began before the fall of 2016 and the work was funded by USAID, not NIH. In the "Understanding the Risks of Bat Coronavirus Emergence" proposal, Daszak clearly states:

> We have begun to characterize the species composition of free-ranging bat populations and have collected samples from over 1000 bat individuals (28 spp.) from 35 localities in over 15 (two-thirds of all) Chinese Provinces. We will also utilize archived wild bat, rodent, and civet samples collected by our

team in Malaysia, Thailand and Indonesia on another large federally-funded [*sic*] project to provide samples of species regularly imported into China (section C1b) (21, 74).

Next, the collected samples are transported from the research site to the laboratory. Once the samples arrive at the laboratory, the samples are immediately stored in deep freezers, are refrigerated, or are immediately tested.

This is dependent on the type of sample (blood, saliva, or feces), the method of testing or analysis to be performed on the biological sample, and how quickly the tests need to be performed (i.e., some biological agents and samples can begin to decay quickly).

The first three assays performed when looking for viruses in a biological sample are often virus isolation, cultivation, and identification. Once viruses are identified in the sample, they can be further replicated to create additional virions for future experimentation, if necessary.

Next, the viruses are sequenced. Sequencing technology has rapidly advanced since 2010. The Centers for Disease Control and Prevention (CDC) defines genomic sequencing thusly: "Scientists use a process called genomic sequencing to decipher the genetic material found in an organism or virus. Sequences from specimens can be compared to help scientists track the spread of a virus, how it is changing, and how those changes may affect public health."[61]

Sequencing can be used to: 1) characterize a virus; 2) estimate a virus or variant's prevalence in the population; 3) potentially evaluate the effectiveness of new medical treatments; 4) investigate future disease outbreaks and disease transmission in a population over time; 5) track the evolution of viruses over time; 6) develop new specific vaccines or medical counter countermeasures to the virus; and 7) edit the virus's DNA or RNA. In the process of viral sequencing, DNA or RNA is extracted from the virus.

Next, the RNA is converted from single-stranded DNA into double-stranded DNA if needed. Then, the strands are chopped into shorter pieces to get a desired length. Then, the ends of the fragments are modified so they can be recognized by the sequencer (i.e., a highly sophisticated and very expensive piece of laboratory equipment, which is typically only owned by research universities, biotech companies, reference laboratories, top tier medical providers, or by state or government owned biolaboratories).

At this point, the sample is called a "library" and is ready for sequencing. Next, the library is loaded into a sequencer, which will identify the nucleotide bases in the DNA fragments (i.e., [A & T] or [C & G]. Lastly, the sequencer produces data—millions of long strings of letters—which are then assembled or aligned with a high-quality reference sequence.

Analytical computer software programs compare the new sequence data to the reference sequence and identify variations in the sample that enables data analysts, bioengineers, and scientists to infer an ancestral relationship, like a family tree.

Everything related to sampling and sequencing that you just read in the previous few paragraphs is contained within one sentence within the Specific Aims.

At this point you might be wondering what happens with all these data. These genetic sequence data are highly valuable for many reasons, but simply put, genetic sequence data are the basis for numerous highly profitable biotechnologies. They are so valuable that numerous people have attempted to patent the naturally occurring genetic sequences.

The US Supreme Court ruled in 2013 that "naturally occurring" genes cannot be patented because they are a "product of nature," meaning that they cannot be claimed as a human invention.[62] However, the US Supreme Court also permitted patents based on laboratory reconstructions of DNA, known as complementary DNAs, or cDNAs.

These sequences and their characterization are the foundation of highly profitable biotechnology, including viral gain of function research and development and medical countermeasures to viruses. These sequences were stored in a database owned solely by EcoHealth Alliance while I worked there from 2014–2016; they were shared between partners like the WIV and at UNC and were also uploaded to a US government-owned database called Genbank.

Sometimes, these samples or the analyses derived from the work were requested by project sponsors at the DoD.

The collection of biological samples containing SARS-CoVs, the sequencing of SARS-CoVs, and the detailed characterization of SARS-CoVs is a critical component of gain of function research and is discussed in detail in Specific Aims 2, but mostly in Specific Aims 3 of the proposal submitted by Peter Daszak.

In Specific Aim 3, he essentially describes all the steps of gain of function, which I will translate: "We will test our models of host range (i.e. [sic] emergence potential) experimentally using reverse genetics, pseudovirus and receptor binding assays, and virus infection experiments in cell culture and humanized mice."

Peter really is communicating: we will make up a pseudoscientific definition and measurement system for something we just invented from thin
air, and which cannot be replicated, called "emergence potential."

Then, we will use genetic engineering (a.k.a. reverse genetics)* to make
SARS viruses that can only replicate once and we will begin to explore com-
pounds that will rapidly examine the selectivity and affinity of a test agent
at a receptor target, so that we can identify a chemical structure to generate
a lead medical treatment for the engineered virus (a.k.a. receptor binding
assays). After that we'll see how well our engineered viruses and pseudoviruses infect human cells (in mice).

 Next Peter piles on the horse manure a little thicker but he does it so
eloquently: "With bat-CoVs that we've isolated or sequenced, and using live
virus or pseudovirus infection in cells of different origin or expressing different receptor molecules, we will assess potential for each isolated virus and
those with receptor binding site sequence, to spill over."

What Peter is really saying here is: using the SARS-CoVs isolated or
engineered with varying types of receptors, we will test all permutations of
naturally occurring and engineered viruses or pseudoviruses to specifically
identify which ones would be the worst for humanity.

Then, we will take the results of this analysis and combine them with
our pseudoscientific measurement system to make many bold and unsub-
stantiated claims about the probability of disease transmission between bats
located in isolated caves, viruses which we just created in the laboratory, and
humans. He then goes on to say: "We will do this by sequencing the spike
(or other receptor binding/fusion) protein genes from all our bat-CoVs,
creating mutants to identify how significantly each would need to evolve
to use ACE2, CD26/DPP4 (MERS-CoV receptor) or other potential CoV
receptors."

This can be deciphered as Peter saying: But first, we must sequence
all the spike proteins (the part of the virus that enables coronaviruses to
penetrate cell walls in animal species) that were found in naturally occurring coronaviruses and the coronaviruses spike proteins that we created to
determine which one has the best ability to infect human cells. Then, Peter
continues: "We will then use receptor-mutant pseudovirus binding assays,
in vitro studies in bat, primate, human and other species' cell lines, and
with humanized mice where particularly interesting viruses are identified

---

* Reverse genetics is an experimental molecular genetics technique that enables researchers to elucidate gene function
  by examining changes to phenotypes (of cells or organisms) caused by genetically engineering specific nucleic acid
  sequences (within DNA or RNA).

phylogenetically, or isolated." This scientific jargon means: We will create binding assays for the engineered (mutant) pseudoviruses that we created. Then, we will conduct a series of experiments in a laboratory where we attempt to bind all the viruses that we possess, both engineered and natural, to a wide variety of species cells, including humans, to determine which ones are the most dangerous to humans. The final statement from Peter is a real gem: "These tests will provide public health-relevant data, and also iteratively improve our predictive model to better target bat species and CoVs during our field studies to obtain bat-CoV strains of the greatest interest for understanding the mechanisms of cross-species transmission."

The final translation: these experiments will characterize how to best engineer SARS-COVs to make them highly transmissible between animals and humans.

There is no possible way to target bats during sampling because this violates the assumptions of predictive modeling (i.e., random sampling), and despite that flaw, past sample collections are not necessarily predictive of future sample collections. A bat captured for sample collection is either a host for a novel (new) coronavirus or it is not.

You know nothing about the probability of sampling a novel coronavirus from previous measurements.

This was my major critique of the proposal then, and I privately told Peter that this project amounted to pseudoscience in its claims.

Members of the executive team at EcoHealth Alliance, Kevin Olival, Jon Epstein, and Peter Daszak, discussed this proposal openly, and there was not any question as to whether this was gain of function work.

It was obvious to all of us, and Peter is obviously an excellent writer.

Despite everything you just read, Dr. Anthony Fauci claimed at a congressional hearing in July 2021 that " . . . I have never lied before the Congress . . ." Which was stated regarding his previous statement that "NIH (NIAID) did not fund gain of function work . . ."

Despite NIH admitting that it had funded GoF at the WIV via EcoHealth Alliance.[63] [64]

After writing this chapter, I later discovered something else: I examined the metadata of the PDF file that Dr. Daszak asked me to review, and it was edited on April 15, 2014. That is interesting and is highly suspicious because the date of April 15, 2014 is after the proposal was originally submitted, according to official NIH date and time stamps, and was during the comment and review period for Dual Use Research and Concern (DURC) in the US Federal Registry where the official US policy for this research was being debated and established.[65] [66]

Was Dr. Daszak provided with insider information from someone at NIH or from Dr. Anthony Fauci?

Did someone at NIH ask EcoHealth revise their proposal submission?

Why was the proposal revised after the submission?

CHAPTER 10

EcoHealth Alliance's Cast of Characters

The people working at EcoHealth Alliance were quite interesting. The purpose of this chapter is to aid any future investigations related to EcoHealth so that investigators do not have to investigate blindly. In my opinion, most of the employees at EHA are useful idiots. I do not mean that they are actual idiots.

Many, if not most, of the people that work at EcoHealth are brilliant.

They are useful idiots because they are the people that genuinely want to make the world a better and healthier place.

They simply don't understand that they are a company that greenwashes the special interests of notoriously dirty industries and harmful large corporations, while they are collecting intelligence for the US government.

Peter himself is caught in a paradox, truly loving wildlife and animals, but loving power and money more.

Determining what everyone did at EcoHealth Alliance was difficult, as there was not any on-boarding process, corporate structure, or processes in place. I identified this as one of the weaknesses at EHA and recommended to Peter that we hire a chief operating officer or give someone that responsibility as we were rapidly growing as a company, or so it seemed.

As a company rapidly grows, standard operating procedures are a necessary component of the business to ensure stability with decreased oversight and insight from managers and executives alike. I quickly realized that Peter didn't want any rules or processes so he could just make up the rules as he went. After he made it clear to me in a passive manner that he did not value process, I would sit back and watch him make countless leadership errors from the safety of my own projects and sources of funding.

One of the stranger management decisions that he had implemented was that there were not any clear salary bands associated with job titles or experience. Peter used this unstructured system to give employees meaningless titles. Unless you were an executive, you were a peon in Peter's management world, and he would negotiate salaries independently with junior employees, which created huge differences in pay across the company.

This was terrible because it created friction and animosity across the company. Apparently, after I was promoted after a short period of time to vice president, Peter also felt the need to promote Dr. Kevin Olival to vice president since he had been with the company throughout his education and scientific career.

I believe Kevin was very likely worthy of the promotion, as I viewed him to be a wise and astute scientist. He worked long hours, was a prolific scientific writer, and had a deep understanding of genetics and evolution in infectious diseases, especially bats and viruses. I think that Kevin felt the need to prove himself to me for some odd reason, but perhaps my perception was off. In one of these moments where I sensed him competing with me, he stopped by my office, which had a large window overlooking the Associated Press building in Manhattan, to discuss the latest in biotechnology, a nanopore CRISPR (Clustered Regularly Interspaced Short Palindromic Repeats) knockout kit which could be licensed for about $3,000 per year. These types of genetic engineering tools use a CRISPR sequence of DNA and its associated protein to edit the base pairs of a gene.[67]

Specifically, Kevin suggested that we could potentially use this new CRISPR kit to quickly splice coronaviruses into their smaller parts, and I agreed with his idea and assessment. Heck, I wanted one just to try it out and to play with it.

The administration of the organization was handled by a sea of mostly youthful administrators. Megan was our office manager who always had a positive attitude and was a fierce negotiator within the office. Joe worked in finance and worked closely with the principal investigators to ensure that we were spending our budgets consistently.

By working with Joe, I sensed that the organization relied on the overhead on personnel to stay afloat. Finances were tight and if we did not bill predictably, the organization ran the risk of running out of cash.

Peter required two administrators. Dr. Aleksei Chamura was attending graduate school to obtain his PhD and served as EHA's chief of staff. Peter treated Aleksei terribly, and I felt disgusted when Peter would scream at him across the entire floor of the office, call him names, and treat him inhumanely, which was an all-too-often occurrence. His other executive assistant, Alison, had more junior administrative responsibilities, and often bore the brunt of Peter's fragile ego and outbursts.

I felt bad for her, too. Alison was highly capable, always friendly, and incredibly well organized.

Dr. Carlos Zambrana-Torrelio was a younger master's degree level scientist while I worked at EHA, and he earned his PhD after I departed the organization. He was an absolute machine in terms of his productivity and had a firm grasp on infectious disease ecology, research design, methods, and quantitative analytics.

Carlos and I worked together on the Modeling and Analytics Team for the USAID PREDICT program.[68] [69] He also understood plant and animal agriculture and the unique relationships between emerging infectious diseases and food production. He and I developed ideas together on how to model emerging infectious disease risk for food and agriculture, most of which did not ever come to fruition mainly due to Peter's inability to understand what we were communicating to him. Carlos played a critical role in helping EcoHealth maintain its standing as a reputable scientific organization.

Dr. Parviez Hosseini was a young PhD, and brilliant mathematician, and I thoroughly enjoyed listening to him eloquently explain complex mathematical models and simulations. He was EHA's quantitative lead on the PREDICT program until he departed EHA in 2015. Parviez built excellent and accurate mathematical models and never overstated his models' predictive abilities.

I sensed that he didn't like where the organization was headed and Peter's frantic leadership style and the chaos he created in the office. Parviez left EHA to work at the US State Department as a strategic information advisor in programs related to HIV and AIDS.

Dr. Noam Ross was hired as Parviez's replacement. Noam was kind and had very strong computer programming skills to match his quantitative abilities. He was a fresh graduate and fell into the category of being an ivory tower academic. I am not sure if he had ever worked at a real job outside of the academy, but I sensed that he had not.

The models that EHA built related to the PREDICT program fell into the category of being built to demonstrate that EHA's programs had value to our project sponsors, which is a very common problem with models (unintentionally and intentionally).

The type of work that Noam was executing for Peter was work that gives statisticians and modelers a bad name by using complex mathematics and computer software to generate the bullshit "right answers" camouflaged in mathematical language relatively very few people can understand.

Parviez was often asked to generate similar bullshit modeling products to prove what Peter had essentially ordered the company to do was working, but he had the business acumen and social skills necessary to say no without saying no to Peter.

Initially, Noam was a huge pain in my ass because he had the same tendency as Peter to distract my employees by requesting and persuading them to work on other projects, which amounted to more timecard fraud against the US government. I suspect that Peter suggested to Noam that he use my employees for his projects as I had previously told Peter to stop the behavior. Nevertheless, I identified the problem quickly and chewed him out in my office for what I viewed as essentially ripping off the government and distracting my employees, which resulted in the behavior change that I wanted. He avoided my staff and me for the remainder of my time while I worked at EHA.

Dr. Catherine Machalaba was earning her PhD in environmental and planetary health sciences at CUNY and is an expert in infectious disease policy. Catherine worked for Dr. Billy Karesh, an EHA executive vice president working in policy, and was able to stay out of much of the fray and fallout associated with Peter's poor leadership since she reported directly to Billy. She frequently traveled domestically and internationally and is very well connected to the most powerful governmental and non-governmental organizations globally.

Dr. Melinda Rostal is a veterinarian, epidemiologist, and field biologist at EHA. Mindy and I frequently chatted when she was not conducting field research on emerging infectious diseases like Rift Valley Fever or Crimean-Congo Hemorrhagic Fever in Africa. Mindy and I both attended the School of Public Health at the University of Minnesota and were both affiliated with faculty from College of Veterinary Medicine that worked in zoonotic emerging infectious diseases (i.e., diseases that are transmitted between animals and humas, like COVID). Mindy also reported to Dr. Billy Karesh when I worked at EcoHealth, although she spent a significant amount, if not most, of her time in the field abroad collecting biological samples and collecting data related to human behavior and risk factors. She has what is considered the dream job by many people that love emerging infectious disease research. She is an incredibly hard worker, which is a common characteristic of Minnesotans.

Dr. Jon Epstein hails from Boston and is a veterinarian and epidemiologist. Jon has a long history of working with bats and coronaviruses dating back to the early 2000s. He has been a pioneer in the field, and I was somewhat starstruck when I began to work with him since I had read and studied much of his work on emerging infectious diseases.

He was earning his PhD in epidemiology and sometimes he would ask me for help or sanity checks with his doctoral research. The behavior of stopping by and asking a person for quick analytical help is a characteristic of people that are trained as clinicians first and scientists second.

Throughout my career, clinicians have contacted me or interrupted me while at work to get quick answers to their research analysis midway through their projects without me understanding the nuance or context of their specific problem. Often, I would be unable to provide the specific quick answer he sought. Often in science, the devil is in the details and not understanding the context can result in the clinician receiving bad advice.

The rumor at the organization was that Jon was being groomed to one day replace Peter, if EcoHealth somehow survives the COVID engineering disaster.

Mr. Anthony Ramos was the communications director and reported to Peter. Anthony and I quickly became friends and remain so until this day. Anthony and I would periodically get together outside of work for drinks or nights on the town with our significant others. Anthony educated me on New York City night life. He was also a classically trained chef, ate out on a regular basis, and knew everything happening in New York.

Anthony served as the corporate liaison to the board of directors at EcoHealth Alliance and could quickly identify a person's personality type and know how to socialize, and was excellent at schmoozing wealthy donors, private foundations, and the New York City's powerful elite.

By proxy, he was one of them, and is a class act.

Anthony was forced out of the organization by Peter when the cash flow dried up after my departure from the company. I never asked Anthony specifically what Peter said to him, but I can only imagine it was another pathological lie to paint Anthony as a failure in some way. Peter never took responsibility for his failures as a leader and blamed everyone else for the organization's problems, but never himself.

By coincidence, Anthony and his husband live near me, and we periodically get together for drinks or dinner.

Dr. Simon Anthony was our resident biological laboratory expert. Simon had a joint appointment with EcoHealth Alliance and Columbia University. He had an excellent sense of humor and handled conflict and Peter's aggressive style in stride. Simon was fortunate that his biological laboratory was at Columbia and that he was typically only at the EHA office for meetings.

Simon worked very closely with Dr. Ian Lipkin at Columbia University, and I believe that Dr. Lipkin had also trained Dr. Jon Epstein on laboratory methods earlier in his career. Ian and Simon worked closely together on identifying and extracting novel viruses from biological samples collected on the PREDICT program. Every virus discovered was an easy peer-reviewed publication and another bullet point for their CVs, and they were excellent at their work.

Christopher Allen (Toph) was a transplant like Peter. He'd lived in the United Kingdom with his father for an extended period, and his mother was from Wisconsin, as I recall. Toph had worked and trained under Peter while earning his Master's in Public Health at Columbia University. Toph reminded me a lot of myself, and we instantly hit it off. He had the unique ability to envision creative solutions to complex problems and understood the importance of providing practical real-world solutions to stakeholders. He was never scared of any technical problem and firmly believed that we could develop technologies to solve many of the most difficult problems in health care and data and analytics.

Toph had been working full time for Peter on other projects, and I quickly realized that Toph was being underpaid and was undervalued by Peter. So, I did what any smart executive would do: I offered him his fair market value in salary and more independence to put his creativity and brilliance to better use in big data and infectious disease technology research and development. He accepted my offer to switch to the department at EcoHealth, and I sold the idea to Peter by appealing to his weakness, money.

By moving Toph to my department, I would cover his salary; Toph agreed to occasionally assist with modeling and analytics on the PREDICT program. The key to winning any negotiation is to make everyone feel like a winner, regardless of any other aspect of the deal.

In my department, I had several rockstars that I hired and a few sub-performing employees that I inherited. The employees that were sub performing were not sub performing because they lacked scientific or technical ability; they were unable to perform because they believed that EcoHealth Alliance was in the business of conservation and improving health.

Peter or Nico had previously recruited a few idealistic useful idiots that would argue with management about the type of work that they felt was impactful or important, instead of executing the work that we were under contract to perform for our customer, DTRA. I made life so process- and defense-focused that those employees naturally left the organization. Peter was happy about this since he wouldn't have to pay them severances or unemployment benefits. I personally would have rather just professionally terminated them.

One of the best employees that I hired has remained my friend and business partner until this day, Brock Arnold.

Brock was a full stack software engineer and had worked at the biggest companies in the tech world. He hated large corporate politics and merely wanted to be able to work from home for a competitive salary that I could offer based on cost-of-living differences, which is why I favored recruiting remote employees over local software engineers working at Google or Facebook. Brock was excellent at writing production code, but he could also effectively manage, train, and mentor junior software engineers, which is a difficult quality to find in the tech world. Brock and I had developed plans to commercially monetize several of the technologies that we developed for the DoD and other agencies.

We were going to accomplish this by creating a new company partially held by EcoHealth Alliance.

I had socialized the plan with members of the board of directors at EcoHealth Alliance, and they were extremely excited about the value proposition. In fact, several other people in my position made millions of dollars by commercializing infectious disease surveillance and modeling platforms. One example is Dr. John Brownstein, who made millions when he sold his disease mapping platform called HealthMap which to a large beltway bandit corporation.

John, and HealthMap, was subcontractor on my large Defense Threat Reduction Agency (DTRA) contract, and our program officer, Dr. Chris Kiley at DTRA, ordered the removal of HealthMap after I requested a modest increase in HealthMap's subaward to the tune of $1 million, which is only a nickel in Pentagon bucks.

Dr. Brownstein was quite upset at this and blamed me for his removal. I am not sure what more I could have done to keep HealthMap on the contract as the decision had been made before I received the call ordering his company's removal. I later learned from Anthony Ramos that Dr. Nico Preston, the former scientist leading my department, finally came to the same realization as Dr. Brownstein and I had about monetizing our scientific and technical work. Nico wanted to spin his department's technical work off into a new company, and Peter balked at the idea.

So, Nico left his position at EcoHealth and walked away from his large DTRA contract, which is a huge deal to any young scientist that lands a big research contract.

I also socialized the idea of monetizing the technology to Peter, who privately told me that it was a great idea and he supported it.

Since I had verbal approval from some members of the board of directors and from Peter, I began to socialize the new company with my staff. We made logos for the company, and I developed the business model and plan, only to have Peter reverse his position publicly about a month later, which I believe he did to humiliate me and undermine my leadership of my department.

This was the first and last time that I had been "Petty Petered."

* * *

Eventually, after I left the company, Brock was put in charge of the department with a tight leash held by Peter.

After I left the company, I did what any smart technologist would do with ideas my company was not interested in: Brock and I founded a new company outside of EHA in only five minutes and started developing new ideas, technologies, and products, which we eventually sold to customers.

I would periodically hear Brock's complaints about work and the funniest was that Peter fell for another email phishing scam which cost the EcoHealth Alliance $25,000. I used "another" because similar hacks to EcoHealth's bank accounts had happened due to Peter's failure to protect the company from even the lowest level of fraud.

And then there is Dr. Peter Daszak: The king of thinking that he is great at business and research and is terrible at business and doesn't understand science, at least in my opinion when I worked with him. Since he was running a scam of convincing everyone that they were doing good for the world, there were no non-disclosure agreements, intellectual property transfer agreements, or non-compete agreements to sign as a condition of employment at EHA. There were no standard operating procedures, pay scales corresponding with job titles; there was no physical security for data or collected samples; and there was no information security or enterprise security for corporate computing.

Peter didn't know what he didn't know, and he used a corporation without rules to abuse his position of authority and his employees. After the first and last time Peter lied to me about privatizing my department outside the company, I decided to start looking for work the next day.

Never work for a person who lies to you.

It can only end badly, and leaving EHA was the best career decision that I ever made.

CHAPTER 11

The Only Scientist at EHA to Ever PREDICT Anything

In February 2015, Dr. Peter Daszak started subtly implying in jokes that I would be let go if I were unable to bring in any large contracts. He was not joking, and I watched him engage in this behavior with other employees.

Thankfully, a few months later I successfully negotiated a new $4.6 million contract with DTRA for a continuation and expansion of my work.

Over the next two months, I rapidly grew the department from four to twelve employees and took on additional unpaid research assistants and interns from Columbia University, where I held an adjunct professorship in Ecology and Evolutionary Biology.

Being an adjunct professor was unfortunately necessary as Peter encouraged the tactic among employees holding advanced degrees so that EcoHealth Alliance would not have to pay for subscription fees to scientific publishers.

I also used the huge influx of funding to upgrade computing equipment and technology companywide, which was a good business decision and a good political decision, as I did not want to create a culture of haves and have nots within the company. We also upgraded our systems architecture for big data and systems availability, and began to build, test, and deploy our technology platforms in the same manner as big tech companies. This massive achievement of my career pulled EcoHealth Alliance from the red and pushed it into the black.

Following this huge influx of cash, Harvey Kasdan, the chief financial officer (CFO), leapt from his chair, screamed out in joy, and gave me a massive bear hug. Additionally, Peter immediately requested a meeting with me to discuss the award notice.

At the meeting, Peter promoted me from senior scientist to associate vice president. I also requested a one-time bonus and a pay increase, which were granted.

I also requested to be added to the PREDICT program. Peter quickly stated that he would add me to the Modeling and Analytics Team on PREDICT and that I could also serve as a country coordinator to foreign countries in the PREDICT program, as EcoHealth, Metabiota, UC Davis, and USAID were in the planning phases of the next large award. This ended up being a real-life lesson in learning to be careful what you ask for. Initially, I skimmed thousands of pages of historical PREDICT work and read in depth any material or peer-reviewed publications related to my responsibilities.

Initially, I wanted to work more on PREDICT than my own portfolio of work, as the hunt for emerging infectious diseases and attempting to predict the locations of their specific emergence was sexy. In fact, saying that it was sexy is an understatement of the importance of successfully being able to predict infectious disease emergence. Being able to successfully predict when and where a specific disease would emerge would likely result in a Nobel Prize at a minimum, and the potential for saving millions of lives at a maximum, and I eagerly awaited my opportunity to successfully do it.

The stated goal of one of the platforms that I was developing was to use something called Natural Language Processing (NLP) to read and process communications from textual sources with computers. The idea was that if we could develop an NLP platform to read and process communications related to infectious diseases, we could then use machine learning (a.k.a. Artificial Intelligence or AI, even though the term AI is technically being misused to market machine learning) to classify emerging infectious disease threats in textual sources.

This technology could then be applied to the pipeline of digital information moving around the globe to process and identify emerging infectious disease threats in near real time. This would be much faster and less costly, and would provide an earlier warning than laboratory-based surveillance methods like antigen, antibody, or PCR tests.

The name of the platform was called GRITS and we also made an advanced programming interface (API) to enable other applications to communicate with our platform. The GRITS platform was fairly accurate in its ability to identify diseases in the massive streams of data, and I can't help but wonder if the DoD used it to detect the emergence of COVID. [70]

The value of this tool is that it could be used by the government to identify potential events to be flagged for review by analysts or could be used to detect a spatial cluster of events; this use of machine learning is known as assisted AI, and we developed it in 2015.

The dirty secret that nobody understood was that by the time someone with enough medical training wrote about the disease in a way that would be classifiable by a computer, the person infected with the disease had likely already been seen by a doctor, which would be a less accurate diagnosis compared to a laboratory test and would only be potentially slightly faster than the traditional laboratory-based surveillance systems.

Seeing the writing on the wall, I began to deprioritize this hold-over application envisioned from my predecessor Dr. Preston. Government program managers and officers can be hit or miss, meaning that some are very passionate about their portfolio of work, while others are merely punching the government clock and are wasting taxpayer money with their mere appearance at work. Dr. Christopher Kiley, our program officer at DTRA, was somewhere in the middle: generally pleasant to work with and have as a customer.

Dr. Kiley liked to check the boxes, and I was happy to make that easy for him to do.

I was far more excited about the other models and platforms that I had already made or envisioned. While leaving Sandia National Laboratories, I was wrapping up a publication on a model that Dr. Nicholas Kelly, Walt Beyler, and a rockstar math intern named Joe McNitt had developed, which discovered that pandemics could cause food shortages based on examining the impacts that labor shortages had on inputs to food production (i.e., water, chemicals, energy, sanitation, food ingredients, and transportation systems) and the overall effect that labor shortages had on the food supply chain itself.[71]

We found that pandemics, when labor shortages are pronounced, can cause food shortages.

This model was validated during the COVID pandemic when food shortages first started to impact the meat packing industry in 2020, which then motivated President Trump to use the Defense Production Act to force meat processors' employees back to work while reducing liability for the meat processing companies.[72]

Peter insisted that EcoHealth take credit for the work, which he had his administrative staff handle, which is technically academic fraud.

Thanks to Anthony Ramos, I was interviewed by the *New York Times*, while I was still working at EcoHealth, to discuss the findings, which were published in the periodical a few days later. [73]

Another idea of mine was to correct a terribly envisioned and executed project by Peter which he named "Sicki-GRITS," which was a half-baked scheme to predict infectious diseases from the historical record of emerging infectious diseases.

This idea was so inherently stupid that I thought he was joking when he told me about the project.

Peter didn't understand prediction or modeling. So, I simply decided to create something of value from this harebrained scheme, rather than tell him that it could not be done or was not a good idea, and then allow him to be the champion of the new idea, and it worked perfectly.

The solution to Peter's problem, which became my problem, was to create an indexed and searchable database called Emerging Infectious Disease Repository (EIDR: pronounced like eider, the species of duck); EIDR Connect to reconstruct disease events across different corpuses of documents; EpiTator to annotate epidemiology data using NLP; FLIRT (Flight Risk Tracker), which was a tool to predict the movement and spread of infectious diseases globally using spatiotemporal flight data for every traveler in the world; and MANTLE, which was planned to be a platform to use NLP to automate the combination of disparate infectious disease databases using ontologies.[74]

My biggest success was FLIRT.[75]

FLIRT accurately predicted the spread of the Zika Virus globally in 2015 and 2016. I was so confident in FLIRT's ability that I contacted a science reporter working with *The Guardian* and provided FLIRT's prediction about where cases would arrive in the United States and at which rates, and the prediction was exactly right.[76] After the success, the only thing Peter could muster was that it would have been better if I published it in the *New York Times*.

I owe much of my success as a modeler to my biostatistics advisor and mentor Dr. James (Jim) Hodges. Jim taught me about how to build useful models. Prior to becoming a professor, Dr. Hodges worked at the RAND Corporation in its heyday where it was his job to build models and evaluate existing models' utility for the DoD. He wrote an excellent paper on how models should be used and how they can be used appropriately, and I had the benefit of him explaining to me how, when, and why to use a model.

If government and corporate executives merely read and understood this single paper written by Dr. Hodges, then the government would save millions of taxpayer dollars and corporations would have save millions in unwasted revenue by knowing when to build a model and how to use it properly.[77]

My goal as a technologist and scientist is to provide value to humanity; unfortunately scientists increasingly have little say or influence in determining what is of value scientifically, as most, if not all, research's value is determined by the organization or person providing the funding and is not determined by the potential value of future accidental scientific discoveries (the most impactful discoveries in human history happened by accident and were not intended or hypothesized from inception).

From Dr. Hodges's wisdom that he passed along to me while he was my dissertation statistics advisor, I have been able to successfully avoid the pitfalls of quantitative modeling, which resulted in me being the only person at EHA to accurately predict anything while I worked there.

If the larger program funded by USAID named PREDICT worked, then EcoHealth Alliance should be famous for preventing pandemics, not causing them.

The USAID PREDICT program was a classic case of overpromising and underdelivering, which I would soon understand in detail.

CHAPTER 12

$$$ Follow the Money $$$

A key characteristic of any for-profit or nonprofit business is how the entity generates revenue. EcoHealth Alliance was a non-profit organization registered in the State of Massachusetts.[78] Prior to 2010, EHA used to be a company called the Wildlife Trust Inc. and, according to my understanding, used to be involved in animal and environmental conservation efforts.

Dr. Peter Daszak transformed the organization by concocting a marketing message that communicated to stakeholders that EHA was engaged in conservation work by protecting the health of wildlife.

While this message has face validity scientifically, very little, if any, work was being conducted on preserving the health of wildlife while I was employed at EcoHealth Alliance.

The work at EcoHealth Alliance mostly consisted of a series of perpetually underfunded scientific research projects which focused on understanding zoonotic disease transmission.

For example, I assumed control of a project where EcoHealth Alliance was paid $25,000 by the United States Department of Agriculture (USDA) to develop and maintain a global spatial biosurveillance system for Ranavirus, which primarily impacts amphibians, reptiles, and fish. That amount of money for software development and platform maintenance costs was grossly insufficient and was the kind of terrible deal only an organization in financial distress would agree to complete. While measuring infectious diseases impacting any species is a worthwhile endeavor, the activity of biosurveillance alone does nothing to prevent the transmission of a disease, to improve the health of animals or humans, or to increase the health of a species' population.

While EcoHealth Alliance was receiving millions of dollars from US government agencies to conduct global biosurveillance, none or very little of the work resulted in reducing the burden of disease or improving ecological health.

This was the snake oil Dr. Daszak was selling to billionaires, global governments, academics, and to conservationists.

From my experience at EHA, as a consultant I always ask potential customers how they generate revenue and how much revenue they are generating. As a potential employee interviewing for full-time employment, I always ask about the financial health of the company. I do this because I do not want to work under false pretenses, and an organization cannot improve unless the employees and executives understand what the financial situation of the company is. A few months after beginning to work at EHA, I began to sense something was off financially based on the look and feel of the office and how business operations were managed. Much of the computing equipment and software systems were out of date and past repair by the manufacturer, all the offices and cubicles had dingy furniture that had been pieced together, the break area was ill conceived and cheap, the conference rooms were tacky and had dysfunctional audio-visual equipment, and there was insufficient space for a growing company. When there were budgetary issues, Aleksei would shuffle office to office, striking deals with staff to move money between accounts. At first, I didn't think anything of the money shuffling because this type of shuffling happens in academic institutions, and I had experienced it before. It wasn't until my promotion that I began to think the business operations were strange, as I saw the company's revenue streams and profit margins for the first time in weekly executive and financial meetings.

Once I began to attend these meetings, I saw the first strange financial aspect of the company. The first was that EHA's indirect cost rate negotiated with the US government was comparatively low compared to other academic institutions, yet EHA was operating a business in one of the most expensive places in the United States, New York City.

Indirect costs are the costs associated with operating a business that cannot be directly billed. When the negotiated indirect cost (IDC) rate with the government is too low, it can greatly reduce the profitability for organizations that operate efficiently, below the negotiated IDC. The other problem with an indirect cost rate that is too low is that operations that protect the business and the employees cannot be afforded (e.g., safety training, security training, equal opportunity training, and ethics training).

Peter's strategy was to use the low indirect cost rate to undercut competitors, bids on government contracts to increase the likelihood of receiving the funding or award. If EHA was successful in the bid, then EHA would have to do more with less as the IDC rate negotiated was insufficient to cover costs on some projects. The only problem with this approach is that a project or program that is not profitable results in an unsustainable business.

I pointed this problem out to Peter when EHA's renegotiated indirect cost rate came in at the 30–40 percent range, as I recall. He was not happy that I pointed this problem out in an executive team meeting.

It was clear to me why EHA was perpetually playing the money shuffle game. For a place of business in NYC, the indirect cost rate should be above 60 percent. The money to remain solvent had to come from somewhere.

EHA also obtained funding, material support, or data from private foundations (e.g., The Wellcome Trust, the Google Foundation, Rockefeller Foundation, and the Skoll Foundation). Most had either donated money to EcoHealth, or in my experience and supported by documents from EHA that I have in my possession, had contracted, or provided earmarked funds, material support, or data, for specific contract work or extensions or formed strategic partnerships with existing US government sponsored programs.[79] [80]

This could be an act of fraud if EHA engaged in these behaviors and the dual streams of funding were not reported to the US agencies funding the work. The practice of billing the US government on a time and materials contract for a project and receiving funding for the same or even partially related work from a different funding source or government agency is known as double dipping.

At a minimum, it is a grey area that is not tightly regulated, which Peter aimed to exploit financially, and the organization's survival was dependent on these types of shady business dealings.

The contracts or donations from private foundations were not enough to keep EHA above water, so Peter would be forced to find other more profitable sources of funding. This is where private donors came into play.

Every year EcoHealth would hold a charity fundraising event at Guastavino's located at the very south end of the Upper East Side of New York City. Peter and Anthony would arrange for exotic animals to be on display so that ultra-wealthy donors could have their pictures taken with the cute fuzzy animals. This was another form of snake oil orchestrated by Peter.

None of the animals on display had anything to do with the work that we were doing at EHA, but they certainly gave the junior staff at EHA and our wealthy donors the impression that we were out saving their cute little lives. The entire charade of the exotic animals at the annual event was emblematic of Peter and EcoHealth.

Attending the event were the members and significant others of some of the world's oldest and wealthiest families and individuals. Also in attendance were people that I would classify as D list celebrities/personalities, photographers, writers, and artists, as well as powerful C suite executives, lawyers, and politicians.

At the annual charity event that I attended;[81] I was asked to court male billionaires that were gay by Anthony.

I am a straight man—not that it matters—but I have been told by gay men, including my friend and former colleague Anthony Ramos, that my fashionable dress attire, personal grooming, and overall appearance causes gay men to find me attractive.

I did not mind being told to dedicate my time with these men at the gala, and I took the request as a compliment. Anthony knew how to throw an excellent party and knew how to work the room full of the global elite. I wanted the company to be successful, so I executed his request.

My charm led to at least one of these men donating at least $20,000 (and potentially almost $200,000 as I don't recall the exact amount) to the organization. It was these types of unrestricted donations that kept the company afloat despite Peter's poor business acumen.

To Peter's credit, he did turn the failing Wildlife Trust into a profitable organization relying heavily upon private donations. Peter's success, in my opinion, should be mostly attributed to Dr. Billy Karesh's abilities and connections, his tough and scrappy administrative staff, and to Peter's unscrupulous practices of both underpaying employees and offering junior employees unpaid volunteer internships.

The wealthy donors that provided access to other streams of money from large multinational corporations, provided professional services gratis, or routinely contributed to the company were offered seats on the board of directors.

The board of directors at EHA was comprised of mostly former or current corporate executives, those from old money families, or those who provided professional expertise to the company. I found it odd at the time that EcoHealth had so many people on its board of directors but have since learned that this practice is common among nonprofit corporations.

If you take a quick look at the current EHA board of directors and search for their names on an internet search engine, you will find that many of them are part of the ultra-wealthy elite, and many have strong ties to the petrochemical, pharmaceutical, agricultural, and food production industries.[82]

Some of these companies have long histories of polluting and destroying the planet; namely putting profits over people. You shouldn't use Google to search due to their involvement in funding EcoHealth, and others have suggested that their search algorithm has been manipulated to information tied to the COVID origin story.

Some of these wealthy people were Democrats and some were Republicans.

Regardless of party affiliation, I believe that their main concerns in life were maintaining or increasing their wealth.

It is only with these powerful and ultrawealthy donors that EcoHealth had enough funding to remain solvent and create SARS-CoV-2 over a period of roughly six years.

From 2014 to 2016, EcoHealth Alliance finally figured out how to obtain large US government contracts.

I had learned how to obtain large government contracts myself from my past experience as a US government contracting officer's technical representative, and from my experience writing, obtaining, and managing complex research and engineering projects throughout my academic career.

Also, obtaining grants for a few million dollars is much easier from the DoD or from a national security related agency as the average contract sizes and awards are much larger in these areas of US government spending.[83]

A few million dollars is a tiny contract in this space, and that is one of the reasons I always targeted these agencies for funding, along with my interest in protecting our country. Peter's tactic of being cheap paid off, but this is not the main reason why Peter was successful in obtaining funds.

Peter created a system to influence the entire field of emerging infectious diseases and scientific discourse.

Peter sat on the grant review panel for the jointly funded Ecology of Emerging Infectious Diseases research panel funded by the National Science Foundation and the National Institutes of Health.

Peter also sat on the board of, or participated in, a long list of scientific organizations as an expert in emerging infectious diseases and was elected as a member of the National Academy of Medicine.

How ironic, given his involvement in causing one of the largest pandemics in history.

If you attempted to have a scientific conversation with Peter, you might be perplexed as to how he was appointed to all these committees and organizations.

But here is how he did it: Peter created his own scientific journal, *EcoHealth*, and is the editor-in-chief of it.

Being a journal editor is a big deal as a scientist or academic, and it enables you to control what the official narrative is within the journal, and (if successful) across the entire field. Simply, Peter would only select or approve "peer-reviewed" articles to be published if they supported his motives and agenda.

Peter would edit the content of the articles or communicate with authors—or have one of his underlings do it—to manipulate the statements of others' work or their scientific findings so that they supported his narrative and funding agenda.

Alternatively, Peter and his henchmen would recommend that authors reference, cite, or quote previously published articles, which would increase the impact of the narrative and would artificially drive up something known as the "impact factor," which is a metric that is used by academics to measure the "importance" or "impact" of their publications.

Manipulating other's statements or work is against the rules in academia, and many scientists have been punished for manipulating and gaming the system. However, there is no way to punish or hold the journal's owner and editor in chief accountable for engaging in these behaviors.

Peter engaged in these behaviors when reviewing my work, my staff's work, and the work of other scientists external to EHA.

Once Peter was able to gain control of the emerging infectious disease narrative and simultaneously held seats on powerful review boards, he could help direct the sea of government funding to people and projects that supported his scientific narrative. This behavior has the net effect of silencing dissenting opinions and prioritizing work that supports his agenda.

Once EHA had secured large US government contracts and subawards, Peter was able to then directly fund subcontractors that supported his narrative, and he had the contractual right to review their work. Since Peter was controlling the narrative of emerging infectious disease science, it was easy for him to get the people under his thumb to make statements related to the specific needs of future research, which typically happens in the last few paragraphs of a peer-reviewed journal.

This was effective at doing two different things: 1) scientists would communicate to other scientists and US government research sponsors that future research should focus on Peter's research objectives; and 2) influencing, via lobbying, the requests for proposals (RFPs) and requests for applications (RFAs) issued by the US government agencies or non-governmental organizations.

And there is more . . .

To put the icing on the cake, Peter, the other executives, scientists, and myself would travel internationally and domestically to present EHA's work to government, corporate, and special interest stakeholders and at academic conferences alike. We would participate in government committees and policy discussions, when and if possible, as subject matter experts.

That way we could influence any potential calls for proposals and would collect business intelligence on future emerging business opportunities in emerging infectious diseases. When all of this was a success, Anthony Ramos would put out press releases with mainstream media outlets. Lastly, we would host private cocktail parties and dinners where our US government project sponsors from USAID, NIH, or DoD would speak, which were attended by diplomats, the rich, the powerful, and high-powered scientists.

The US government employees never accepted gifts and even paid for their own dinners, but that doesn't mean that our influence peddling and narrative-controlling machine was not effective.

We were highly effective, and EHA even contemplated an office in Washington, DC, for the sole purpose of more effectively steering funding to our company.

This is how you run a completely corrupt scientific organization.

This is the perfect example of the financial, ethical, and moral corruption in academia and collaborating organizations, and these behaviors also occur at many of the leading research universities in the United States.

Peter understood exactly what he was doing, and his excellent writing ability paid himself and EHA dividends in his ability to shape the narrative. In fact, what Peter was executing was exactly what I was trained to do as a PhD student from my Department of Homeland Security Research Center of Excellence directors Dr. Amy Kirch, Dr. Frank Busta, and Shaun Kennedy and mentor Col. John Hoffman.

However, I never thought to entirely control an academic journal to shape the narrative. That part was brilliant.

Even when you have completely shaped the scientific narrative to your advantage, things may not go your way. While EHA dramatically increased its winning percentage on government contracts and grants, EHA still lost out on contract awards. But I had learned another better approach in business: having leverage during the negotiation.

Peter spent a great deal of time with Anthony shaping his and EHA's image instead of focusing on making the business profitable, and this was one of his fatal flaws as a leader.

In my opinion, Peter always put himself before his employees and the company, and he wanted to be *perceived* as a great scientist instead of *being* a great scientist.

CHAPTER 13

The $210 Million USAID PREDICT Fraud

Many of the statements that I make in this chapter have been recorded on video under penalty of perjury or have been submitted in writing as sworn declarations under penalty of perjury.

PREDICT, a project of USAID's Emerging Pandemic Threats (EPT) program, was initiated in 2009 to strengthen global capacity for detection and discovery of zoonotic viruses with pandemic potential. Those include coronaviruses, the family to which SARS and MERS belong; paramyxoviruses, like Nipah virus; influenza viruses; and filoviruses, like the ebolavirus. PREDICT has made significant contributions to strengthening global surveillance and laboratory diagnostic capabilities for new and known viruses.

Now working with partners in 31 countries, PREDICT is continuing to build platforms for disease surveillance and for identifying and monitoring pathogens that can be shared between animals and people. Using the One Health approach, the project is investigating the behaviors, practices, and ecological and biological factors driving disease emergence, transmission, and spread. Through these efforts, PREDICT will improve global disease recognition and begin to develop strategies and policy recommendations to minimize pandemic risk.[84]

After skimming and reading thousands of pages of PREDICT reports and funded publications, the project appeared to be the solution to emerging infectious disease threats, and I wanted to be a part of it.

The program and research described previously sounds amazing, does it not?

That was the problem: everything sounded amazing, but once you looked at the details of the projects, the entire program amounted to pseudoscience.

So, what was PREDICT, really?

After being promoted and assigned to the PREDICT program on the modeling and analytics team and as a country coordinator on the hunt for MERS coronaviruses in Africa and the Middle East, my conversations with unrelated DoD officials took a different tone. Increasingly, I was being asked about the infectious disease data and analyses that we were collecting in Southeast Asia and the Middle East.

Specifically, the DoD was asking me to hand over EHA proprietary data and analyses related to Peter's Hotspots program, which I would soon determine was not valuable since the predictions were likely to be inaccurate based on flawed assumptions and data in the model, and the genetic sequences that we were collecting.

I did not find the questions to be strange because I was keenly aware why the DoD and the DTRA would want to have and see these data.

The DoD and DTRA would first want to evaluate these data, models, and analyses to see if they could predict infectious disease risk using something called a Technology Readiness Level or a Military Readiness Level to determine whether the technology should be invested in or used in military biological threat and risk assessments.

Since I quickly inferred that that was what DoD, DTRA, and national security agencies wanted from EHA, I successfully used EHA's proprietary work as bargaining chips in large contract negotiations. I am assuming that the national security and intelligence community eventually came to the same conclusions that I had come to regarding the very limited predictive power of PREDICT's models as we were only able use to use a fraction of the data once in contract negotiations.

Never give the government anything of value unless you must contractually.

This is how you play hard ball with the federal government. Government employees always want to buy a new Ferrari and have it delivered in twenty minutes, while the government can only realistically afford a used, high-mileage PT Cruiser that needs some work.

I perceived Peter's primary objective related to PREDICT to be always focused on sample collection, viral isolation, and storage of genetic information. Peter frequently dictated to all staff members the importance of storing all sequenced data in a database which EcoHealth Alliance owned and maintained named Eidith.

At the time, I didn't quite understand why Peter was so interested in the genetic sequence data as they were not impactful to PREDICT's stated objective of predicting and preventing pandemics. Infectious diseases do not exist in a vacuum, and they are competing against each other for survival and dominance in a limited population of disease reservoirs. To determine which diseases are threats, we first need to understand the transmission risk from animals to humans, and then we need to know what the risk is for sustained human transmission.

A simple fact is that the bats being sampled to isolate the coronaviruses that they carry did not pose a significant risk to humans because humans were rarely exposed to the bats located in caves in remote areas globally, especially when compared to the human exposure risk of other infectious diseases. Despite this obvious flaw in the logic behind the PREDICT program, Peter always ensured that all samples collected in the PREDICT program and across all other projects where biological samples were collected were entered into the Eidith database.

From personal communication with members of Decentralized Radical Autonomous Search Team Investigating COVID-19 (a.k.a. DRASTIC),[85] EcoHealth and UC Davis are now conveniently directing investigators of the origin of COVID to USAID. I bet Peter, Kevin, Jon, or EHA has retained a copy of these data, and they should be questioned under oath with penalty of perjury related to all biological and viral samples collected at EcoHealth Alliance.

These biological and viral samples collected across programs, which were funded and enabled by wealthy donors, board members, private foundations, foreign governments, and the US government agencies, were combined and could be used for the gain of function work directly funded by Dr. Anthony Fauci at NIH in GenBank[86] and individual sequences can be searched for similarity with the Basic Local Alignment Search Tool (BLAST).[87]

This was apparent to Peter, Billy, Jon, Kevin, and me, and would be apparent to any scientist that worked in this space.

Data is the gold that scientists collect, and it will be used in any way that generates new publications and results in additional funding. Data is the currency of research and science, and for Peter and EHA's Eidith database was used to store valuable genetic sequence and disease data (the ownership of Eidith was later assigned to USAID PREDICT and is demonstrated in the development history on github).[88] [89]

The executive team openly discussed the gain of function work related to the NIH proposal, as any scientific research team would do naturally, and Peter was insistent that all sequence data from PREDICT were entered into Eidith and GenBank as fast as possible. These were the data and samples that were foundational and were used to construct SARS-CoV-2 in the gain of function research, and this fact was described in the proposal "Understanding the Risk of Bat Coronavirus Emergence."

There were even discussions led by Peter about how they could leverage the work and biological samples across projects, which is also another tactic among successful scientists. Specifically, there were discussions as to how the PREDICT samples could be used in the NIH gain of function work, which made perfect business sense. The most profitable, and often most impactful, research is the previous research that you can get more mileage out of by repurposing and repackaging as new.

When the discussions related to gain of function came up during executive meetings, I made my opposition to the work clear, and this might have been why I was excluded from the NIH project (thank God), but more likely that Peter was trying to keep the costs down in the proposal.

Over the course of several months, there were numerous and endless PREDICT planning and coordination meetings. The PREDICT core partners UC Davis, Metabiota, and EHA were negotiating over the countries that each partner would assume the responsibility for and the funding that accompanied the contracts.

Naturally, the partner that had the strongest ties to each country would typically negotiate fiercely for the country to be included in their portion of the project and budget. Peter wanted to control the globe, and he would do or say anything to capture as many countries as possible, which was good for him personally and for EHA.

As a new country coordinator, I discussed as part of a group, with Peter leading the discussion, which countries would be included in the budget, and I was assigned to South Sudan and Jordan upon my request, due to my experience working with people from northern and western Africa and my time spent working in the Middle East.

In Jordan, I was to be paired with Billy, who had previous experience working with the Jordanian government.

In South Sudan, I was likely to be paired with Peter, but the funding from South Sudan was eventually reallocated due to armed conflict in the region.

When the conversation came to China, I made my opposition to working in China clear, and I think Peter took this as a slight against him and EHA's work as Peter viewed the work in China as the most important to the company. I could not and would not work with a communist country that had been stealing our country's and American corporations' intellectual property and technology, and I viewed scientific engagements with the Chinese as a significant national security threat to the United States, and potentially detrimental to my ability to obtain another top-secret security clearance.

At a minimum, it would have created significantly more paperwork and would have been a hassle to obtain a security clearance due to the reporting requirements. At this meeting, I felt the need to openly question EHA's involvement with the Chinese as I was concerned that the Chinese would steal our intellectual property, deceive our organization, or lie directly or by omission, which are well-documented behaviors of the Chinese government and the companies and scientific organizations that are controlled by their government and the communist party.

After I raised these concerns, there was a silence among the executive team, until Peter decided to quickly dismiss my concerns.

After countries were assigned among Peter, Billy, Jon, Kevin, and I, we began building our research teams abroad. We needed to identify personnel to collect bats in the field and find laboratories to analyze and store the samples. Simultaneously, we began meeting and constantly communicating with USAID program managers and the US State Department Missions stationed at the Embassies in our host countries.

We were expected to take calls at any hour of the day based on their 9-5 schedules, which is what good government contractors should do to support them. The deputies of the embassy wanted to know every aspect of who we were speaking with, who we were contracting with, where we intended on collecting samples, and who would analyze these samples.

I had never experienced such terrible micromanagement in my life as a scientist, and I couldn't understand why any of these people were so heavily involved in our operations. Operationalizing, sample collections, and analysis is not difficult work, and their micromanagement was irritating and insulting. Nonetheless, I was polite and answered their questions and did what they asked.

Once I received the budgets for Jordan and Sudan, I was shocked at how small they were. After EHA took its overhead for managing the contract, there was barely any money to execute the work. As an American company, we benefitted from the exchange rate when working in poorer countries. If the USD is significantly higher than the local currency, then we could negotiate subcontracts which were lucrative for employees in the host country and were a cost savings to our organizations.

Even with the foreign cost advantage, there was not enough money to collect and analyze enough samples to have sufficient statistical power to be predictive of anything.

This was the moment that I realized that USAID PREDICT and EcoHealth were a complete farce.

This coupled with the strange micromanagement from Dr. Andrew Clements at USAID and other senior embassy personnel (a.k.a. office foreign missions or diplomatic missions), I began to sense that EcoHealth Alliance was really in the business of collecting intelligence on foreign laboratories and personnel, while on the hunt for coronaviruses.

I didn't have a complete grasp of what I was experiencing, but I knew something was off.

As a part of the project planning process, Billy and I planned to go to Jordan on a scouting trip, and we arranged meetings with the leaders of Jordanian health and agriculture-related ministries, university leaders, and professors, where we would evaluate the capabilities of potential new partners and cement new relationships.

The endeavor was highly political and sensitive to the US government, and we took the trip very seriously.

Billy and I prepared, and I was to watch Billy handle the first few engagements with powerful stakeholders. I already knew that Dr. Billy Karesh was an excellent statesman and listening to him to speak and engage with foreign leaders was one of the best learning experiences of my life. After Billy handled the first few engagements, I began to take the lead over the discussions and Billy would smoothly inject himself into the conversation when I faltered.

By the end of the first day, I had mastered the engagements, so Billy was able to sit and smile as I introduced and discussed our scripted programmatic goals and objectives and secured our partners' verbal commitment to the project. While engaged with these foreign dignitaries, I felt like a diplomat, and I felt proud of what we were trying to accomplish and to be improving American foreign relations.

These meetings, independent of the "work" that we were claiming to do, were probably the most important aspect of the PREDICT program—foreign diplomacy and building good will, where we could, to prevent biological threats. Bioweapons don't kill people, people kill people, and people, including myself, frequently make mistakes.

I found some things problematic during the trip to Jordan. As part of our partnership building process, we would be taken on tours of facilities including their biological laboratories, none of which met American environmental health and safety standards and had outdated equipment.

Billy and I would look at each other while touring these facilities and we knew exactly what the other person was thinking—these labs were grossly inadequate. In the appropriate context, the people running these laboratories were doing the best they could with limited resources and hopefully US government funding could improve the environmental health and safety of these laboratories.

My main concern was the physical security of the samples including deadly viruses that would be collected in a remote bat cave that would be transported to a laboratory, stored, and analyzed. In academia, in the United States and abroad, physical security of laboratories is all too often insufficient.

At the end of our trip, we spent an afternoon touring Jordan and looking at old Roman castles with a quick visit to Petra. Upon our return, Billy ran into several other US government contractors and personnel to whom he introduced me, and we all shared a bottle of wine and talked about research, our work, and international diplomatic issues.

The next day I returned to New York City.

Back at the EHA office, I was behind on projects related to my portfolio of technology development, but thankfully Brock was able to lead and hold down the fort with Toph in my absence.

At the next PREDICT meeting with the executive staff, I voiced my concerns over laboratory biosafety and biosecurity, and with the biosafety and biosecurity awareness and training of the people that were executing this dangerous work for EHA overseas. PREDICT and its partners were required to complete various levels of biosafety and biosecurity training, which in my opinion were inadequate for the type of work our foreign partners were performing. This was later validated by journalists imbedded with EHA staff members while sampling bats for coronaviruses.[90]

Additionally, training is only as effective as a supervisors' ability to evaluate laboratory and field workers' biosafety and biosecurity behaviors and performance, while using any identified deficiencies, incidents, or accidents as opportunities to train and mentor employees. There was no such framework in place at the laboratories that I had visited nor were there effective biorisk, biosafety, and biorisk management systems or frameworks in place to supervise or mitigate the risks associated with foreign laboratories on the PREDICT program, which was a massive liability risk to EHA. Peter did not care and dismissed my concerns as he deemed the existing policies effective.

Despite PREDICT's problems, I enjoyed the diplomatic nature of the work. Who wouldn't enjoy a swanky scientific job traveling the planet, acting as a diplomat, and making deals? Despite my short-term excitement, my conscience was weighing more heavily on me. Especially when I sat in the PREDICT modeling and analytics team meeting where Peter would ask Noam and Kevin to perform nonsensical modeling tasks that were time intensive and amounted to snake oil.

I was increasingly becoming quiet in meetings, and this, combined with Peter's other terrible behaviors, motivated me to find work elsewhere.

When the government offers cash for programs like these, charlatans and snake oil salesman thrive. Government reports are rarely peer-reviewed, and there is no requirement that the scientific work being performed must perfectly match the government reports, and if they are required to match, the government program managers rarely read the scientific publications to ensure that they support the conclusions in the reports.

Most troubling, there is no requirement that the scientific or engineering work being performed is accurately interpreted. In fact, the US government, including the US military, project managers and officers are allowed by regulation to rewrite and interpret findings, and some of the work that I did at Sandia National Laboratories was rewritten to fit the narrative that our government program sponsors sought.

This fact has been well-documented, and the ultimate example was the rewriting and interpretation of the development of the Bradley Fighting Vehicle, a US government boondoggle that cost billions of dollars to develop a troop transport vehicle that couldn't transport soldiers.[91] The USAID PREDICT program's official reports reek of these types of manipulations as the methods and data collected could not support the grandiose claims made in the reports.

My Experience with and Opinion of Dr. Ralph Baric

The single biggest failure of the PREDICT program, which resulted in the development of SAR-CoV-2, was the introduction of Dr. Shi Zhengli to Dr. Ralph Baric by Peter. Dr. Baric is a micro-biologist at the UNC Gillings School of Global Public Health.

The Gillings School of Global Public Health is ranked third in NIH funding,[92] and Dr. Anthony Fauci has been its de facto Don for decades. Ralph specializes in gain of function (GoF) research and adamantly believes without any scientific justification that GoF experiments can make a virus more deadly and transmittable and that many GoF experiments are worthwhile.

I was introduced to Dr. Baric at EHA where he gave a scientific talk during lunch. His presentation was so dull that I tried my hardest not to fall asleep. He walked us through his gain of function methods and presented them in the context of how they could be used to predict emergence risk, and I was not buying it.

At the end of his talk, I asked a few questions specifically related to his methods and assumptions, which supported the claim that they could predict emerging infectious disease risks from the gain of function work described in the NIH proposal, while Peter made a sneering expression in my direction.

Baric responded by making an argument to his authority and experience, while not addressing my question specifically. I did not press him any further as I believed that it was a waste of my time. Ralph famously later stated: "You can engineer a virus without leaving any trace. The answers you are looking for, however, can only be found in the archives of the Wuhan laboratory."[93]

Dr. Baric developed numerous methods and techniques at UNC, which make him the grandfather of gain of function research. Dr. Baric increased the virulence and transmissibility and infectivity of viruses by cutting their RNA into strips and altering each individual piece of genetic material.

Ralph also created "no-see-um" viruses, through passaging, which were designed to intentionally hide the evidence of genetic manipulation.[94] Baric stated that the no-see-um method had "broad and largely unappreciated molecular biology applications."

Ralph also developed methods to recreate the genetics (RNA & DNA) of eradicated diseases through reverse genetic engineering. Dr. Baric had also mastered recombinant experiments which can use the cut pieces of genetic material from several different diseases to manufacture artificial contagious viruses.

At UNC, Ralph used and further developed a technique known as passaging. Passaging is the practice of injecting a live virus into an animal or adding it to a cell culture in the hope that the virus will adapt to the new species. After the virus is added to the animal or cell culture, the virus is removed after a period before it is no longer able to replicate within the animal or culture and is then transferred to another animal or cell of the same type.

This is often an iterative process.

Some types of viruses mutate quickly and, through the process of passaging, will adapt quickly to the new host, which results in a new and more deadly pathogen. Ralph also successfully cloned the entire full-length infectious mouse-hepatitis genome, and his "infectious construct" replicated identically to nature.

The specific targets used by Baric were "humanized mice" and human cells grown in cultures. Humanized mice are genetically engineered to carry a human protein named ACE2 which covers the surface of cells that surround human lung tissue.[95]

These techniques and others were taught to Dr. Shi Zhengli thanks to PREDICT's funding establishing the necessary relationships in China, the next round of supporting funding from NIH, EHA's private donors, and by Peter introducing Ralph to Shi. Ralph taught Dr. Zhengli the best existing methods to engineer bat coronaviruses to attack other species.[96] [97]

This type of information exchange is called "technology transfer" by the US government, and this instance specifically was a transfer of the most advanced GoF methods in the world from a US-sponsored laboratory to the Chinese via EHA, and the only person that seemed to have an issue with transferring dual use GoF data and biotechnology to the Chinese was me.

My Experiences with and Opinion of Dr. Shi Zhengli

I briefly met Dr. Zhengli in passing at EHA when she made a few appearances at the office when I worked there. I barely recall her scientific talk at EHA, and like Ralph's, it was rather boring and not persuasive. As part of PREDICT's snake oil development and marketing program, they funded the collection of viruses found in bat blood, feces, and urine. They greatly overhyped that if these viruses "spilled-over" from nature directly or through an intermediate species to humans, they could cause the next pandemic.

An additional part of the PREDICT data and sample sharing agreement with the Chinese, the samples would be stored in the BSL-2 lab, which is typically an appropriate level of biosecurity for dealing with foodborne pathogens like *Salmonella Spcs.* at the WIV. This was not the appropriate level of biosecurity for dealing with viruses which Peter and PREDICT collaborators claimed could cause the next pandemic.

In collaboration with the Chinese, PREDICT, EHA, and its partners collected more than twenty thousand bat samples and identified roughly two hundred coronaviruses which were generically like SARS-CoV-1. Shi was the virologist in China at the WIV that led sample collection and analysis from samples collected in China.

While the mainstream media treated her as some great scientist, a more accurate characterization of her would be that she was a novice virologist compared to Ralph.

Ralph handed Shi all his cookbooks and mentored her on viral GoF research and development, and she either rapidly became one of the best virologists in the world or was a complete moron akin to your local methamphetamine dealer attempting to cook meth for the first time and then blowing up their own house and all their neighbor's houses on the first attempt. Personally, I find the latter hard to believe.

With the USAID PREDICT funds, Peter developed a nonsense investment pitch which was communicated to everyone, including Shi and Ralph, that we needed to conduct GoF experiments to see which bat coronavirus was likely to spillover from bats so that we could develop a vaccine for it before it became a pandemic.[98]

In 2019, as PREDICT was winding down, Peter tweeted this nonsense argument to further persuade his followers that his work was important and that it would lead to vaccines and therapeutics for a laboratory engineered coronavirus, which did not even exist in nature:

> We've made great progress with bat SARS-related CoVs, ID'ing 50 novel strains, sequencing Spike Protein genes, ID'ing ones that bind to human cells, using recombinant viruses, humanized mice to see SARS-like signs, and showing some don't respond to MAbs, vaccines we've now shown that people highly exposed to bats have antibodies to these bat SARSr-CoVs, and are working in those communities to reduce risk this [*sic*] might not yet be business as usual, and it might not yet work for other viral groups. but it's proof-of-concept in a v. important viral family with pandemic potential all of [*sic*] this work is published, with more on the way and a bunch of papers from

Ralph Baric's lab [*sic*] Here's the critical point – Ralph's work shows, for some or [*sic*] our newly discovered SARSr-CoVs, putative SARS vaccine and MAb therapeutics don't work. Viral discovery, therefore, is ID'ing CoVs related to a known pandemic strain, but that are able to evade vaccines/therapeutics![99]

CHAPTER 14

Mob Boss: The Psychopath of Infectious Disease Pseudoscience

In the winter of 2015, my department at EHA was a well-oiled machine. I was presenting my work globally and domestically, and it was being well-received everywhere it was presented. Yet, I wanted to leave the organization as quickly as possible because I felt like eventually everyone that funded EHA would see it for what it was, and I was naïve at the time to think that the project sponsors didn't already know what EHA really was.

So simultaneously, I was interviewing for executive level positions at corporations and for professorships at universities.

I had several offers for employment before February 2016, and I was indecisive for one of the first times in my life, as I was hoping the craziness at EHA would subside.

In late 2015, around the holidays, Peter and I both stayed late after work. It was common for all the executives to work late into the evening. That night I had ordered an early dinner and ate at my desk while I completed a proposal that I was writing which was due in a few days.

At roughly 9:30 p.m., I decided to call it a day, packed my briefcase, put on my jacket and gloves, and began to head out the door. As I approached the front door, Peter came out of his office and asked me if I was headed home too, which I affirmed.

Peter and I shut off the lights, and I held the main door open for him. As I locked the door, Peter asked me, "Andrew, do you have a minute? There is something that I would like to ask you about."

I responded, "Sure, Peter."

As we stood there facing each other in EHA's dumpy vestibule, he said, "Someone from the CIA approached me. They are interested in the places we're working, the people that we are working with, and the data that we are collecting."

I was stunned. I couldn't believe what I was hearing! I couldn't believe it for a few reasons.

The first was that if you had a security clearance you would never say something this sensitive in public. This conversation would be conducted in something called a secured compartmentalized information facility (SCIF) if the people in the conversation both had security clearances. From this, I garnered that Peter did not have a security clearance or, if he did, he didn't care about the rules.

The second reason was that I was not sure if Peter was being targeted by someone and that conversation was a setup to entrap Peter in something that he probably wouldn't be astute enough to detect.

Any foreign agent could generate fake credentials and approach Peter pretending to be from the CIA.

I had to think quickly as Peter and I were staring each other in the eyes, and I stated, "Peter, it never hurts to talk with them; there could be money in it."

He nodded in agreement, and we got on the next elevator from the seventeenth floor to the lobby. We made small talk about our holiday plans and commuting to work while walking out of the lobby, across the street, and while walking north on 10th Ave for about a half a block, where Peter parked his car in a surface lot.

We said our goodbyes, and I continued my walk north on 10th Ave to my apartment on 45th Street.

When I arrived home, I told Emily about my day at the office, including the conversation with Peter about the CIA. I was excited that it would open new opportunities at the organization, and, knowing Peter, there was a specific reason why he approached me with the news where and when he did.

I didn't know the half of it at the time.

This was one of the last cordial conversations that I had with Peter Daszak. The next month, PREDICT work was really ramping up, and the executive staff were all traveling for one reason or another.

On three separate occasions between meetings, Peter quietly confirmed to me that the work with the CIA was progressing, with no substantive details.

For all I know, Peter could have been lying to me or playing one of his stupid games, but I believed him at the time. While I was still planning on leaving the organization soon, I kept on working hard on PREDICT and the rest of my work. I am a perfectionist, and I was taught in the infantry to always leave a project or place better than you found it.

Also, the reality was that I would still likely have to have conversations with the numerous collaborators of EcoHealth Alliance and wanted to remain on good terms with EHA staff as the world of emerging infectious diseases is very small. At some point in the future, it was very likely that our paths would cross again.

Over the next few weeks, I became more active in the PREDICT modeling and analytics work, and Carlos and I were discussing ways to model emerging infectious disease risk in food and agriculture, an area where I had extensive experience and professional relationships. I was excited to make something that could predict emerging infectious disease risk in agriculture production.

I socialized my ideas for building a predictive model using network models based on production locations, consumption locations, and the emerging infectious disease risks associated with production locations and specific food commodities. Carlos and Kevin liked the idea, and when I socialized the idea with Peter, he thought it was a good idea too.

So, I took the initiative and booked a trip to Rome to meet with the United Nations food and agriculture researchers that were part of the PREDICT modeling and analytics team, and I combined the trip to the UN with a mandatory global PREDICT meeting, which I recall was being held in Dubai.

I told Peter that I was traveling to the UN no less than three times in various meetings over the course of a month, and I was certain that we were all on the same page about the food and agriculture modeling work.

Everything seemed to be going according to plan.

January through February was a difficult period because I had several great employees ask for pay raises, and the spinoff company that I had socialized with some of the board of directors was shot down by Peter unexpectedly and without reason after he'd previously agreed to it. Many of the employees in my company were upset about the news and a few were discussing taking other more lucrative positions at larger tech companies, which I knew that they could probably easily obtain.

My budgets were maxed out, and I had little room to negotiate salary increases. I was also hearing from Anthony, who worked with everyone in the company more regularly than I did, that employees outside of my department were feeling jealous about my department's larger salaries, ability to work from home, and my very flexible time-off policy.

I have always taken being a leader seriously and part of keeping great employees is being a fair and flexible boss to the greatest extent possible. I believe in servant leadership (i.e., empowering your employees to self-improve and become leaders themselves by focusing on their well-being), and I wanted to help them all be champions.

With their happiness and success, I would be successful.

So, I did what any good boss would do, and I requested a meeting with CFO Harvey Kasdan and Peter so that I could argue for a company-wide pay grade system and a solution to increase salaries for my staff, the entire company, and myself. I even offered to create the pay grade system myself to help make salaries and promotions transparent to the staff.

Under the existing system, compensation was a one-on-one negotiation, and promotion and standing within the company were not clear to the staff. Peter handed out titles which had no clear meaning to employees within the company. Outside of my department, everyone worked and reported to Peter as he saw it, and a CEO should not be so engaged in micromanaging and negotiating junior staff salaries and duties. The CEO should be focused on the future of the organization and generating revenue, not the minutiae of daily business and operations.

As I saw it, his inability and inexperience in leading companies was hurting and causing conflict within the organization.

When the meeting was scheduled and finally held on February 10 or February 11, 2016, Harvey, Peter, and Peter's aide-de-camp Aleksei were in attendance. We met in the conference room after normal business hours. Aleksei opened the conversation with the stated goals and objectives, and then I made my case as to why my department needed pay raises and plainly stated that if the employees left, then we would not deliver on our projects on time to the DoD/DTRA.

I provided examples of what our engineers and data scientists were earning and compared them to the leading tech companies, where their compensation would be two to three times higher than the compensation that we could offer at EHA. I also said that EHA needed a formal structure for promotion, employee review, and compensation so that promotion and compensation would be clear to everyone. I talked about how and why this could reduce conflict across the company and in my department.

It only took Peter about two seconds after I finished making my case to quip some petty, glib statement which inferred that I didn't know what I was talking about, that the problem that I proposed was not really a problem, and that everything was fine.

He was delusional.

These comments, coming from the man who screamed at his subordinates across the office like they were unruly children, were quite rich. Before the meeting, I envisioned Peter's potential responses to my comments and concerns. This type of response is what I expected most as our relationship seemed to be rapidly deteriorating as of the end of January. With exception of Dr. Billy Karesh and myself, Peter recruited mostly people with passive and weak personalities or dispositions to work at EHA so he could work as a dictator who would publicly undermine staff that drew his ire.

I was instantly incensed and, in a fit of anger, I responded by calling out all the financial fraud that I had witnessed at the organization over the past year and a half, and Harvey immediately jumped into the conversation to calm Peter and I down.

I then used my own salary as a benchmark and compared it to what people in my position were making at other organizations.

Peter looked at me in disbelief, and then I realized that Peter had no clue what he was doing as a CEO because he didn't have the slightest understanding of our employment market. Harvey insisted that Peter think it over and wanted to call another meeting after my trip to the UN and Dubai.

We ended the meeting with nothing being resolved.

* * *

The next day I had to travel to Rome, and it took me three and a half hours in an Uber to commute from 45th St. and 10th Ave. to John F. Kennedy International Airport.

Upon arrival to Rome in the morning, I hopped into an Uber, went to my hotel in Rome, got cleaned up quickly, put on a suit, and went over to the UN. After clearing security, I was met by our PREDICT partners, who took me on a tour of the building and introduced me to a few stakeholders and collaborators.

After a swift tour and brief conversations, I went to the staff offices where I learned more about their work, and how the UN functions in practice, and then I went into the details of my proposed model for the food and agriculture program.

They loved my idea and my approach, and I stated that I would send them a conceptual diagram and a base prototype model in two to three weeks. After a half days' worth of meetings, I was spending the weekend in Rome before heading to Dubai.

The next day I received a series of calls and text messages from Aleksei and Alison, stating that Peter needed to speak with me. We first arranged calls in the morning, which were canceled, and then rescheduled every one to two hours until 10 p.m. Rome time. Peter had a habit of not respecting other people's time and not keeping scheduled meetings. His staff would frequently change meeting times and would expect that you could cancel whatever other meetings were arranged with other people to accommodate Peter's schedule. However, this was the worst that I had ever experienced. At that moment, I decided to quit EHA when I returned to New York. I turned off my phone and went out drinking and dancing to blow off steam.

The next morning, I decided to go for a walk around Rome, and my cell phone was pickpocketed. While on my walk, I decided to instead quit working on the PREDICT program and work only on my projects to isolate myself from my terrible boss.

I wrote an email to Peter explaining to him that I was quitting the PREDICT program and sent it before I left Italy. I didn't want to have a gap in employment, I had not finalized two of my competing job offers, and I did not know when a viable start date at the new employer of my choice could be.

I canceled my flight to Dubai and returned to NYC instead of attending the global PREDICT meeting.

Upon returning to NYC and walking into the office, everyone was surprised to see me. That is when my research assistant walked into my office and told me the news. Harvey had died of a heart attack on February, 11, 2016, shortly after our contentious meeting.[100]

Harvey was obese and lived an unhealthy lifestyle, so it was unsurprising that a stressful meeting like what occurred likely exacerbated his pre-existing conditions, as I saw it as an epidemiologist.

I informed my staff that I was quitting my work on the PREDICT program to only focus on our department's work. I was terribly saddened by Harvey's death as he and I frequently talked about fishing, which he loved to do, and he had invited me on a future deep sea fishing excursion. Harvey really understood government contracting, and I think he appreciated my managerial style. I made work easy for him.

After I publicly came forward with this story, others suggested that Harvey had been killed, but I have no evidence or reason to believe that claim.

The next few weeks were quiet while the majority of EHA's staff was attending the meeting in Dubai, and I decided where I would work next, fully expecting Peter to act in some petty fashion upon his return. After a few weeks, Peter returned from his travels abroad. When Peter returned, he sent Aleksei over to my office to tell me that Peter wanted to meet with me, and that no date and time had been set. This was Peter's *modus operandi* of attempting to assert his dominance, and I took it in stride. When Friday approached, Aleksei stopped by my office and asked me to meet with Peter late in the day. When I went to the meeting, Peter launched into a spiel about how things weren't working out. That week I learned that he was embarrassed in Dubai when I quit the program and had to explain it to the other PREDICT partners. He then told me that I had to quit. Not that he was firing me.

I launched into an argument pointing out how I was the highest-earning employee and had received a near perfect performance evaluation from him a month earlier. I then told him that I would leave the organization and take the DoD and DTRA contracts with me, and if he attempted to fire me, I would sue EcoHealth for wrongful termination and creating a hostile work environment, which would have been easy to prove. He had nothing of substance to say in response, and he proceeded to call me names. I laughed hysterically at Peter while Aleksei just watched in terror. Aleksei entered the conversation and started to attempt to make a deal. My start date at my new place of employment was set to begin in August, and I wanted to earn enough money so that I could go on an extended vacation in Europe.

So, I made Peter an offer in the spring of 2016: I would resign my position effective June 30, 2016, and I would not attempt to take my contracts with me when I left, which commonly happens in research organizations and academia as scientists move to other institutions. The thing Peter loved more than power was money, and EHA couldn't afford to lose the contracts that I'd been awarded.

Peter grudgingly agreed, and I went home a very happy man after having achieved all my goals.

The next week, or week after, I submitted my resignation letter, which was announced to the staff. At the subsequent executive meeting, Peter put me on the spot and asked me what I was going to do next. He turned bright red and scowled when I told them I had accepted a tenure track professorship offer at Michigan State University.

I later learned from other EHA employees that Peter often manipulated employees into getting them to quit when EHA was low on cash, as if they were being fired. He did this so he would not have to pay them unemployment benefits due to laying them off or firing them without good cause.

This is standard of how Peter conducted business daily.

What an outstanding member of the scientific community!

Just prior to my leaving EcoHealth Alliance, Dr. Billy Karesh seemed to be trying to smooth things over between Peter and I. Billy and I had a great working relationship, and he was incredibly smooth in his communication (a rare quality for scientists, generally speaking). Billy was a veterinarian by training and was an executive vice president at EHA.

This technically made him the second in command to Peter.

Billy had an incredibly impressive CV. He began his career at Plum Island, which used to be the zoonotic infectious disease laboratory off the coast of Plum Island. At Plum Island, a biosafety level 4 laboratory (BSL4), they conducted research and experiments on the nastiest zoonotic agents known to man. After leaving Plum Island, Billy conducted zoonotic infectious disease research globally on a wide variety of species and diseases, and published hundreds of peer-reviewed papers in biological infectious disease fieldwork, infectious disease ecology, and infectious disease policy.

Billy is the true star at EcoHealth Alliance, and I often wonder how it was possible that Peter was running EcoHealth, and not Billy.

I once asked Billy about his desire to lead the organization, and he plainly stated that he liked being in his position. Billy's primary responsibility at EHA was infectious disease policy. He spent much of his time traveling internationally, meeting with the leaders of foreign governments, US government agencies, and leaders of powerful organizations, and courting powerful and ultra-wealthy individuals. He is an incredible spokesman, is a statesman, and can read a person quickly.

Billy also stated that he didn't like having to deal with managing the operations and finance of the organization, and since Peter wanted to have all that power and control, he did not mind letting him have it if he could focus on policy work. His focus on policy paid itself in dividends to the organization, as I believe Billy was responsible for securing much of the funding for EcoHealth Alliance.

Billy and I spent hours strategizing and working on proposals together, and I knew there was much that I could learn from him. During work hours at EHA, he and I were always the first people to arrive at the office, where we would jump back and forth between each other's offices, bouncing ideas off each other and determining how we could leverage existing resources and previous work to obtain new sources of funding.

Billy and I worked on the DTRA's Cooperative Biological Engagement Program[101] (CBEP) proposals together, which were where organizations like EHA basically whored themselves out to larger beltway bandits and contractors to be positioned as sub-contractors in the proposals. The thing that all the potential CBEP partners were most interested in was EHAs massive network of foreign partners (i.e., personnel in foreign countries and foreign laboratories that could execute the biosurveillance and were eligible to receive US government funding to enhance foreign lab capacity).

***

To this day, I still do not understand how building foreign laboratory capacity benefits the United States.

Not understanding the benefit does not mean that we wouldn't pretend like we knew how these programs benefited the United States. To obtain funding, we would merely regurgitate the information that was provided to us in meetings, presentations, or in proposal requests sent to the project sponsor or funding agency. From my perspective, increasing biosafety and biosecurity in foreign laboratories made sense if we had constant and timely visibility into these laboratories, which in my opinion, was not the case.

You might be wondering why I am talking about the CBEP.

The CBEP is the program that funded the biological labs in Ukraine, whose existence the United States government initially denied, and then painfully and awkwardly their existence was admitted by US Under Secretary of State for Political Affairs Victoria Nuland in March 2022.[102 103 104 105 106 107]

EcoHealth Alliance was one of the companies engaged in the CBEP program, and so was their USAID-Emerging Pandemic Threats-PREDICT partner, Metabiota.

Perhaps you may have heard of Metabiota or recall my discussion of it in the previous chapter?

Metabiota is partially owned by a venture capital firm called Rosemont Seneca, and they also received investment from the CIA and DoD venture capital firm In-Q-Tel.[108 109 110]

Rosemont Seneca is partially owned by Hunter Biden, and numerous emails found in his abandoned laptop contain communications related to the backdoor dealings to obtain government contracts responsible for hiring, managing, and funding these foreign laboratories. [111]

I am not sure why the US government needs companies like EHA and Metabiota to be middlemen in managing foreign laboratories for programs which are to help protect the warfighter and the United States. I don't think using middlemen is a judicious use of taxpayer funding, as these middlemen are like leeches that suck the revenue out of the budget and reduce the amount of real work that can be executed in country.

There have been numerous crazy theories circulating about the biological laboratories in Ukraine. The only information that I have seen related to the labs in Ukraine that appears to be worth investigation is the apparent testing of experimental drugs, therapeutics, and vaccines on members of the Ukrainian military and regional population.[112 113 114] It might come as a shock to some of you, but most experimental drugs are tested in randomized controlled trials (RCT) in resource-poor countries where the standard of medical care is beneath that of the United States or Western European nations.[115 116 117 118 119]

The argument that clinicians and scientists like myself make to institutional review boards (IRBs), the organizations which are responsible for evaluating research to determine if they are ethical, is that the standard of care that participants located in poor countries received while enrolled in RCTs is better than what they would receive in their normal daily lives.

From the documents that I read related to the experiments that were being conducted at these Ukrainian laboratories, they seem normal in this context, but there is one area where my perception is that the work being conducted at these labs was somewhat strange.

What was strange was that the specific diseases and drugs being tested were infrequent or rare infectious diseases, which would typically pose an unlikely and low-risk threat to a military force deployed to Ukraine, and that the countermeasures being tested would not pose much value and would be a great expense to stockpile and maintain.

The massive cost and logistical burden of maintaining vast quantities of vaccines for a wide variety of infectious agents is the real reason why the mRNA platform was advocated and selected as a bioweapon, bioterror, and pandemic medical counter measure (MCM), and I will explain the rationale as to why the mRNA platform is a tool which serves a specialized purpose later in this book.

Here is a preview: the mRNA platform was not used as originally intended in response to SARS-CoV-2.

The CBEP was one of the last proposals that I helped develop with Dr. Billy Karesh while I worked at EcoHealth Alliance, and I do not regret formulating the CBEP programs conceptually. At the time when we co-wrote the proposals (2015 and 2016), there was no discussion of testing MCMs, nor was there any discussion of conducting randomized controlled trials in foreign countries. Even if testing of MCMs using RCTs in poor countries was part of our discussions, I might not have objected as this is not a simple black and white issue. The devil is always in the details.

I can infer that the scope of the CBEP contracts changed after I assisted with the proposals, and the budgets were increased after successful implementation, which is a common reward for good performance in government contracting. I still believe that the CBEP and BTRA programs are well-intended, but it is the execution that needs to be improved. There is a fair amount of what appears to be foreign propaganda mixed into Ukrainian US funded biolab discourse. One of these examples includes a DTRA CBEP Biological Threat Reduction Program on past tick exposure, which is hardly nefarious and is well intended, in my opinion.[120] I got the sense that Billy was sad to see me leave the organization. Or, maybe he was happy that I was leaving and he was putting on a great show for me?

Peter, Billy, and I are all great chameleons, which is a common trait among executives and top performing scientists. Billy would contact me every few months after I left the organization while I was working at Michigan State University and at JUUL Labs.

Every six months or so he would ask me to share some of the interesting data that I was collecting with his collaborators or his protégé, Dr. Catherine Machalaba, who was one of the best policy people working in infectious diseases and is a kind and pleasant person with whom to work.

While working at Michigan State University (MSU), I began building two separate portfolios of work. The first was a portfolio of antimicrobial resistance work, a dull but very important area within emerging infectious diseases. As antimicrobials are increasingly used, we select for pathogens which are genetically resistant to the antimicrobials, and, as scientists, we were looking for ways to optimize their use in animal production to preserve their effectiveness in human medicine. This saves animal producers time and money and increases their profitability, while ensuring that these valuable medical tools remain effective to treat infectious diseases, which are increasingly becoming resistant to a very limited set of antimicrobial treatments.

At MSU, I was introduced to Dr. Paul Bartlett, who had previously served in the CDC's Epidemiologic Intelligence Service, which was a prestigious position in epidemiology where PhD level epidemiologists would rapidly respond to and investigate infectious disease outbreaks in animals and people.

I personally no longer think very highly of the CDC based on its numerous failures over the past twenty years. However, agencies change over time, and I suspect that the public health response failure led by Dr. Anthony Fauci and his band of sycophants will result in the CDC being restored as the entity responsible for infectious disease outbreaks, epidemics, and pandemics.

Dr. Paul Bartlett was a very easy going and kind full professor, and I instantly admired his caring approach with his students and his dislike of university administration.

We collaborated on a few projects together where he introduced me to Dr. Marcus Zervos, the chief epidemiologist in the Henry Ford Medical System who is a board-certified infectious disease specialist. Dr. Zervos and I began to strategize several unique ways to backward trace human cases of antimicrobial resistant (AMR) infections back to animal production facilities to determine how much of a role human exposure in animal production contributed to AMR in patients in healthcare (many of the patients with AMR infections had connections to animals, he'd observed). He was one of the best medical doctors and epidemiologists that I had ever worked with, and he really understood the complex nature of exposure pathways and disease resistance, and we immediately hit it off.

In July 2020, Dr. Zervos conducted a retrospective observational cohort study examining the effects of treating patients diagnosed with COVID-19, and mainly found that patients who were diagnosed with COVID-19 and then treated with hydroxychloroquine within twenty-four hours of admission to the hospital reduced the risk of death by about half (26.4 percent to 13% percent).[121]

Shortly after this study was published and publicized, President Donald Trump cited the study, and the fact that hydroxychloroquine is an effective treatment of COVID-19 instantly became heresy among the long list of medical professionals appearing on television that were not worthy of the medical or scientific doctorate degrees that they had somehow received.

My second portfolio of research, using Bluetooth beacons and sophisticated advanced sensor technology to measure and monitor human physiology, became my true passion. I tested this advanced new technology that was pioneered, designed, and built with my business partner, Brock Arnold, after we both left EHA.

Brock is an amazing person who is dedicated to helping his Native American tribe, the Mohawks, improve education and business opportunities for his people. Brock is also one of the brightest software engineers that I have ever met in my life. He is literally capable of doing any software design and engineering task and finds quick and easy solutions to complex problems. Therefore, I'd hired him as a senior software development engineer at EHA, and he'd been promoted to director at EcoHealth Alliance.

After I departed EcoHealth Alliance, I would often listen to gripes about Dr. Peter Daszak from Brock, and I would help him navigate business dealings with my former dishonest, greedy, and egomaniac boss. (Just my opinion!)

CHAPTER 15

COVID Made in America: Bioweapons, Gain of Function, and DURC

Men make history and not the other way around. In periods where there is no leadership, society stands still. Progress occurs when courageous, skillful leaders seize the opportunity to change things for the better.

—President Harry Truman[122]

Biological agents have been used as a military tactic for millennia, and I will skip the roughly first three thousand years of their use, as most bioweapons use during that time falls into the category of poisoning, which is truly unique.

The use of biological agents in warfare has centered on four main strategies: (1) disseminating inoculated blankets or clothing; (2) launching infected materials into enemy positions; (3) inoculating traditional weapons with potentially infective materials; and (4) contaminating food or water supplies.[123]

In the twentieth century, I estimate that over half a billion people have died due to infectious diseases using conservative estimates from the best available data and a simple mathematical model (I used a model, and it could be that their estimates are often inaccurate).[124] [125] The Japanese caused numerous deaths in China by using biological agents and toxins as weapons against them during World War II.[126 127 128]

Over the last 120 years, as our knowledge of the biology of disease-causing agents (a.k.a. pathogens) like viruses, bacteria, fungi, and prions increased, there were legitimate concerns that modified pathogens could be designed and engineered, which would be devastating if used for biological warfare, bioterrorism, and, as I recently realized, as a new form of asymmetrical warfare, which was inspired by the strange emergence of SARS-CoV-2, which is discussed later in this book.

In the 1800s, Louis Pasteur and Robert Koch inadvertently increased the ability to improve and enhance biological weapons because their work enabled humans to select and design biological agents on a logical basis.

This concept of selecting and designing biological agents is emblematic of how we view what is known as gain of function (GoF), the practice of enhancing the pathogenicity, lethality, virulence, or transmissibility of biological agents via serial passage, which is like the methods deployed by Dr. Ralph Baric's SARS-CoV-2 gain of function research and engineering.

Once the potential dangers of biological agents and their manipulation were realized in World War I and later in World War II, that resulted in two international declarations that prohibited the use of biological weapons: (1) in a 1925 Geneva Protocol titled *Protocol for the Prohibition of the Use of Asphyxiating, Poisonous or Other Gases, and of Bacteriological Methods of Warfare*; and (2) the 1972 Biological Weapons Convention (BWC). Both have been international biological weapons policy failures.

While the treaties and declarations prohibiting biological weapons were all made in good faith, they contain no means of enforcement, and they have failed to prevent people from developing and using biological weapons. Any policy, law, or regulation is only as effective as its ability to be enforced. For example, during World War I, biological weapons were used on a small-scale targeting animals and meat production and were not particularly successful.

After the war, false intelligence reports from several European countries created fear, resulting in the creation of several new biological warfare programs before the onset of World War II (WWII).

Sir Frederick Banting, the Nobel-Prize-winning discoverer of insulin, created the first private biological weapon research and development center in 1940 with corporate sponsorship, independently of any state sponsor.[129]

In 1943, the US government was persuaded by their British and French allies to begin a bioweapons program authorized by President Franklin D.

Roosevelt. Research continued following WWII as the United States built up large stockpiles of biological agents and weapons.[130]

Before and during WWII, the Japanese created a largescale biological weapons program and eventually used them brutally against China. During the war, the Japanese army poisoned more than one thousand water wells in Chinese villages to "study" cholera and typhus outbreaks.

The Imperial Japanese Army Air Service dropped plague-infested fleas over Chinese cities and distributed them by hand across the countryside. Some of the epidemics the Japanese caused persisted for years and continued to kill more than thirty thousand people in 1947, two years after the Japanese surrendered to the US.

It is my personal belief that the Chinese people were one of the most negatively impacted by bioweapons due to the illegal and immoral use of biological warfare, and I feel terrible for the immense harm that these biological agents caused to the Chinese people.

While I am certainly not supportive of many of the ruling communist party's international policies and actions in China currently (in October 2022), we all must remember that biological agents do not care about our political leanings, religion, or economic objectives. To use biological agents to intentionally harm others, plants, or animals is to destroy all life, including the people that we love or hold dear, ourselves, and fragile ecosystems.

There is no tactical situation where their use will reach a desired goal, even from the perspective of a rational terrorist who seeks to obtain social dominance through fear, unless the person deploying them is a madman who is willing to kill all life, including their family and themselves. From my personal experience working in national security, people do not understand the black swan (rare but catastrophic) risks associated with their development, research, and use.

Humans have a terrible ability to imagine or quantify black swans.

Humans are wired to ignore these specific types of threats as they are trivial for short term survival, and their mitigation runs counter to the things that bring humans pleasure: our natural desire for freedom and our economic and financial goals. Because of our weaknesses as humans, we believe that we can control things that we simply cannot, and with hubris many people erroneously believe that we can control nature and biological agents.

Although, heuristic thinking is in our nature as humans; history has repeatedly proven that humans will always make mistakes and that nature always wins.

In my experience, the more comfortable a person is with microbiology and biology, the greater their belief that biosafety and biosecurity measures can effectively control infectious disease accidents or escapes from controlled settings. This confidence is normal in humans because they are trained to believe that safety controls are highly effective, which they are, and that regular training and supervision of people that work with agents prevents accidental escapes and transmission of agents, which it does.

The problem is that biological accidents or biocontainment escapes occur within a complex system.

Biosecurity controls are a process of layering pieces of Swiss cheese on top of each other in the actual form of mechanical processes, operational procedures, or environmental health and safety technology. Each biosecurity technology created by humans has technical vulnerabilities, design flaws, or production flaws. Each human implementing biosecurity technologies makes mistakes in their proper deployment and use of the technologies.

Each human conducting operations in accordance with biosafety and biosecurity best practices and procedures makes mistakes.

Thus, biosecurity and biosafety are inherently imperfect and flawed systems.

The goal of security, and, in this case, biosecurity, is to layer these pieces of biosecurity and biosafety Swiss cheese on top of one another so that eventually the holes within each layer are covered by the next subsequent layer of biosecurity and biosafety cheese. The level of biosecurity to be applied to any given environment or agent then becomes a question of cost versus benefits before and after risk mitigation solutions are applied, and this bio risk management process is not concrete or stable over time.

Another way of looking at biorisk management decision making is: What is the risk tolerance for the consequences of an accident when it occurs, compared against the costs of implementing preventive controls, in terms of time and money, to mitigate the identified vulnerabilities or threats which contribute to risk? Vulnerabilities are considered in terms of their probability of failure, which is often quantified by the manufactures of biosafety and biosecurity technologies and is more difficult to quantify in terms of human behavior.

In this case, human behavior could be a person that is known to be bad at performing biosecurity and biosafety operations correctly, which could be quantified based on frequency and severity. Often, there is not this high level of supervision of people that work with biological agents, and the frequency and severity of biosecurity and biosafety mistakes made by personnel are never actually known unless the mistakes result in disease. To add to the inherent difficulty in managing bio risks, past (independent) biosecurity and biosafety failures are not predictive of future failures.

These are all the inherent flaws in biorisk management, and it is not simple or easy to effectively implement.

Most biological laboratories in the United States do not engage in this level of deep thought about how their systems can fail nor do they have to as the agents that people are working with are well understood, the diseases that they cause are easily identifiable, and the treatments to the diseases are readily available.

This is not the case with two classes of infectious diseases: (1) novel pathogens that have been discovered for the first time (i.e., PREDICT's actual work); and (2) gain of function experiments where the result of the next or final step is not known.

EcoHealth Alliance was engaged in both types of work and the bio risk management in place was essentially to trust our foreign partners to implement the proper biosafety and biosecurity technologies, procedures, and training according to our guidelines with very little or no supervision or verification by EHA.

Now, add to this the fact that infectious agents cannot be seen by the naked eye unless they are manufactured in massive bulk quantities and that individual human behaviors and their adherence to correct operational procedure is critical to preventing biological accidents.

With each step there is some residual risk, which cannot be easily measured or eliminated, and no layer of security is ever perfect.

As the number of processes in which the biological agents are handled or manipulated increase, and as the number of preventative control measures increase, the complexity increases.

Often in complex systems, one small failure can trigger a series of cascading failures, which result in the failure of the entire system.

An example of this is the Northeast Blackout of 2003 where a sagging power line on a hot summer day hit tree lines in Ohio, which were then complicated by human error, software issues, and equipment failures. This led to the most widespread blackout in North American history that affected forty-five million people.

In the context of emerging infectious diseases, this type of complex failure occurred in 1977 with the escape of the H1N1 influenza virus in the Soviet Union.

The Soviet Union established Biopreparat, which was a massive biowarfare project that, at its height, employed more than fifty thousand people in various research and production centers, even though they had just signed the Bioweapons Convention. The magnitude of the Soviet Union's efforts were staggering: they produced and stockpiled tons of anthrax bacilli and smallpox virus, some for use in intercontinental ballistic missiles, and engineered multidrug-resistant bacteria and viruses, including plague and H1N1.[131]

At the time, the Soviets were attempting to develop a live attenuated influenza virus (LAIV) vaccine and were simultaneously conducting gain of function research on H1N1 via serial passage. Evidence strongly suggests that the epidemic was caused by vaccine challenge clinical trials for the LAIV, which is characterized as high-risk biological work as described using the framework and criteria outlined previously.

As is the case with SARS-CoV-2, there was no evidence of a natural origin and all the evidence pointed to gain of function research and a vaccine development program gone bad. The only question that remains unanswered: was it part of a weapons program or part of an innocent quest to develop an effective flu vaccine?

Don't fall for my framing of the argument, as the question proposed is stupid.

Infectious agents do not care about our moral justifications or reasoning.

Complex systems like these will always eventually fail, and it is only a matter of when they will fail, how many victims there will be, and how much it will cost humanity for one person's failure.

Even though Drs. Peter Daszak, Anthony Fauci, and Ralph Baric, virologists globally, pharmaceutical companies, and the defense industrial base, along with their cult of brain-dead parroting worshippers, argue that gain of function research helps us to understand how viruses transmit and results in effective vaccines and treatments; to this day, we still do not have a reliably highly effective influenza vaccine, and serial passage gain of function experiments have been ineffective at creating one.

The devil is always in the details. Serial passage live attenuated vaccines—which are used to expose a person to a less virulent strain of a virus while creating a large or sufficient immune response in the person receiving the vaccination, which confers long-term immunity—have been widely successful in protecting people from over twenty diseases. I have personally received most of these vaccines.

This type of gain of function work via serial passage is safe and effective and is not representative of the more advanced and ridiculous work mismanaged by EHA.

To streamline the biosafety and biosecurity management process, bio safety levels were created to quickly apply a reasonable risk benefit calculation for clinical managers, environmental safety officers, hospital epidemiologists, laboratory managers, and government regulators.

These biosafety levels (BSL) are on a scale of one to four, with four being the most controlled safety environment.

Each BSL has a list of biological agents which are appropriate for each level, and it is acceptable to use a higher biosafety level for an agent categorized in a lower biosafety level, but that rarely happens in practice due to time and cost constraints in the context of business. Even government or academic research is a business.

When I was a hospital epidemiologist, I believe that many of the clinicians that I worked with thought I was overzealous in my commitment to biosecurity and biosafety. I once mentioned that the veterinary hospital at MSU was inadequately prepared for a BSL-3 patient, and I suggested training and additional equipment for clinicians to handle potential cases of BSL-3 agents when and if they presented at the hospital during a clinical faculty meeting.

One of the senior clinicians scoffed and rudely retorted, "We don't treat BSL-3 agents at our clinic" and a few of the other clinicians chuckled.

I didn't respond because some fights are not worth having.

The reality for clinicians is that when a sick person or animal presents at the clinic, especially in an emerging infectious disease outbreak scenario, humans and animals present with symptoms—which are often quite similar between different causative biological agents—and the patients have no clue what is causing the disease, and neither do the clinicians initially while they are poking and prodding their patients.

While BSL-3 zoonotic agents are rare in humans and pets, there are occasional cases when they do occur, and it is best to have the training and equipment to properly limit your exposure to the agent and limit the contamination of the clinical environment until the person or animal can be sent or transferred to the appropriate facility.

In the context of SARS-CoV-2, laboratory BSLs will be discussed later in the book.

After World War II, the Soviets convicted some of the Japanese biowarfare researchers for war crimes, but the United States granted freedom to all researchers in exchange for information on their human experiments. These war criminals became respected citizens, and some went on to found pharmaceutical companies.

Soon after WWII, the US military started open-air tests, exposing test animals, human volunteers, and unsuspecting civilians to both pathogenic and non-pathogenic agents. A release of bacteria from naval vessels off the coasts of Virginia and San Francisco infected 800,000 people in the Bay Area. Bacterial aerosols were released at more than two hundred sites, including bus stations and airports.[132]

President Richard Nixon decided to abandon offensive biological weapons research and signed the BWC in 1972.

In 1995, the Aum Shinrikyo cult used Sarin gas in the Tokyo subway, which killed twelve train passengers and injured more than five thousand people. The cult previously attempted on several occasions to distribute (non-infectious) anthrax within the city, with no success, which demonstrated the difficulty in weaponizing biological agents.[133]

The 2001 anthrax attacks, which appear to have been performed by a United States Army Medical Research Institute of Infectious Diseases (USAMRID) vaccinologist named Dr. Bruce Ivin, were highly successful at terrorizing a nation by simply sending letters containing weaponized anthrax spores to specific high-profile people.

In 2002, an update to the Biological Weapons Convention (BWC) mandated compliance verification and failed to garner international support due to the US government suggesting that mandated compliance be dropped and replaced with largely unenforceable provisions which would not mitigate the bioweapon threat. The provisions were investigations of possible non-compliance; assistance in the event of a threat of the use or the actual use of bioweapons; national criminal legislation; the setting up of a Scientific Advisory Panel; revised Confidence Building Measures a new convention on the physical protection of dangerous pathogens; a new convention criminalizing the violation by individuals in the prohibitions of the Chemical Weapons Convention (CWC) and BWC; increased disease surveillance efforts; codes of conduct; and universal membership of the CWC and BWC.[134] [135]

Any nation that signed the BWC with the ability to develop biological weapons, labeled or described as bioweapons, has abstained from doing so. However, the Soviet bioweapons program demonstrates that international treaties have been basically useless unless enforceable verification procedures exist. Even if there were enforceable verification procedures for bioweapons, a country could argue that their biological gain of function research program is being used to develop therapeutics and vaccines against infectious diseases, which also can be used for the development of biological weapons. This is what the essence of Dual Use Research of Concern (DURC) is.

The roots of the concept named Dual Use Research of Concern sprouted with the creation of the USA PATRIOT Act, which was signed into law in 2001, and the Bioterrorism Preparedness Response Act, which was signed into law in 2002.[136] [137] These acts together established the statutory and regulatory basis for preventing the misuse of biological materials within the United States. Specifically, they created the Select Agents List, a designation of restricted individuals that are not permitted to possess select agents, and a regulatory system for the physical security for the most dangerous pathogens in the United States.

In 2004, the National Academy of Science's Committee on Research Standards and Practices to Prevent the Destructive Application of Biotechnology, headed by Dr. Gerald Fink, was charged with minimizing the risks associated with bioweapons proliferation without hindering the progress of biotechnology.[138] The rationale was that biotechnology would be misused by hostile individuals or states.

The implied goal was to accelerate biotechnology development to advance the United States' ability to detect and cure disease.

The stated goal was to make recommendations that achieved an appropriate balance between the pursuit of scientific advances to improve human health, welfare, and national security.

One of the key findings of the Fink Report was that without international consensus and consistent guidelines for overseeing advanced biotechnology, limitations on certain types of research would only impede progress on biomedical research in the United States and undermine our national interests. The Fink Report also recommend the creation of a National Science Advisory Board for Biodefense (NSABB) to provide advice and assessments to the government and the scientific community.

The committee also recommended that the Department of Health and Human Services (DHHS) augment the already established system for review of experiments involving recombinant DNA conducted by the NIH to create a review system for seven classes of experiments involving microbial agents that raise concerns about their potential for misuse, which all support my previously provided definitions of GoF.

What is amazing is that this report emphasized the need to rely on academics and scientists to police and regulate themselves, while repeatedly stating the need for and implying that gain of function work was important and valuable without providing any specific examples of GoF that had been beneficial to public health.

Some of the major concerns and weaknesses identified in the report were poorly trained personnel in biosecurity, biosafety, bio risk management, and physical security. In my opinion, many of the recommendations were not useful.

For example, first imagine your local university, say maybe, the University of North Carolina (UNC).

Next, imagine your resident expert GoF virologist Dr. Ralph Baric is having frequent meetings with the intelligence community and local law enforcement. After having studied for many years at universities and having worked as a professor at universities, I know that research professors running laboratories simply do not have the time to chit chat with law enforcement. What would Ralph and law enforcement officers discuss exactly? His latest bat fecal serial passage humanized mice GoF technique?

Additionally, one of the major criticisms of private industry, which owns and protects most of our nation's critical infrastructures, is that the intelligence community (IC) shares very little information with them, and when information is shared by the IC, it has very little value for operational security.

Even if a professor obtained a security clearance and received a classified warning related to a threat, the professor would not likely be able to communicate the problem to management.

When the IC receives a specific credible threat related to you or your work, they will find you and let you know what to do. From my perspective, the committee members did their best to provide recommendations to implement newly forming DURC policy.

In my opinion, the solutions recommended placed too much of the DURC regulatory burden on the scientists and academic institutions to self-regulate. This is a critical flaw in the overarching bio risk management framework, bioterrorism and biowarfare threat reduction, and DURC policy.

Scientists and universities typically always do the bare minimum to save time and money, and the universities and scientists do not want to impede their own work, which reduces profitability, R&D velocity, and new or continuous sources of funding.

Going back in time, the NIH first formed the Recombinant DNA Advisory Committee (RAC) in 1974 in response to concerns that there was a potential for serious adverse events associated with this new technology.

The RAC was created as a stop gap measure in the absence of regulation supported by legislative action. The RAC was established to provide recommendations to the NIH director and to create a public forum for discussion of the scientific, safety, and ethical issues related to basic and clinical research involving recombinant or synthetic nucleic acid molecules.

In 2019, NIH refocused the RAC into a role closer to its original mandate, which was to follow and provide advice on safety and ethical issues associated with emerging biotechnologies. This is one of the responsible oversight committees that should have reviewed EcoHealth's GoF work.

Institutions that receive NIH support for recombinant DNA research must establish an Institutional Biosafety Committee (IBC) as part of their compliance with the NIH guidelines. While working at MSU, I served on the Biosafety Committee, and I found the extra work to be rewarding. You can probably guess what I am about to say . . . EHA did not have an IBC when I worked there. Nor was there a Biological Safety Officer (BSO) at EHA. BSOs are responsible for laboratory environmental health and safety inspections, as well as risk management of biological laboratories.

They are critical to managing biological risks, especially on GoF and DURC. Since EHA was a passthrough for most laboratory funding, which then went on to external institutions like Columbia University, UNC, or the WIV, EcoHealth would state in their proposals to NIH and other project sponsors that biosecurity, biosafety, and ethical review boards were the responsibility of the contracted laboratories, which, at best, is an unethical business practice as EHA could be held liable for their subcontractors' failures or accidents.

At worst, is a clear and direct violation of federal rule, regulation, or policy related to biosafety and biosecurity at NIH.

A Principal Investigator (PI) in academia is like God. When you become the PI on a large government contract or grant, you have many responsibilities to the project's sponsoring agency, and with great responsibility comes great discretion. PI's have broad discretion in managing government funds and managing the research project so that it is successful.

Typically, most agencies have monthly, quarterly, or annual reporting requirements, and you are evaluated on how well you managed the funds and how well you achieved the specific aims of the proposed work.

These are basic PI tasks, and Peter Daszak had his minions do all this work for him.

When I worked at EHA, I do not ever recall Peter reading official government policies related to our work, nor did he seem to have any grasp of federal contracting, research, or biological safety policies in general. Whenever I mentioned my concerns, he quickly dismissed them or ignored them altogether. Since it is research, project sponsors understand that research projects fail. If we knew what we were doing all the time, we wouldn't call it research. The expectation is that, when an aspect of your project fails, you do the best that you can to quickly create new related research projects and spend any remaining funds on them.

Peter's interpretation of PI broad discretion must be the broadest of anyone at NIH because he didn't believe that he had to follow the rules, nor that his behavior would ever catch up with him. Since Peter was conducting and managing GoF DURC research sponsored by NIH, he had certain obligations and unique responsibilities that he was required to follow.

According to the NIH Grants Policy Statement, and the NIH Consortium Agreement Policy:[139]

> The recipient, as the direct and primary recipient of NIH grant funds, is accountable to NIH for the performance of the project, the appropriate expenditure of grant funds by all parties, applicable reporting requirements, and all other obligations of the recipient, as specified in the NIH Grant Policy Statement.

This means Dr. Peter Daszak and EcoHealth Alliance are responsible for any of the GoF DURC handiwork performed, and any failures, negligence, or accidents committed by the consortium partners (a.k.a. sub-award) like Dr. Shi Zheng Li at WIV and Dr. Ralph Baric at UNC. Whether Ralph engineered the final agent or not, it does not matter. What matters is whether the work that the EHA employees performed while Daszak was the NIH project's PI, or whether the work performed by any of the consortium's partners—Zheng Li (WIV) and Baric at (UNC)—contributed to any part to SARS-CoV-2.

For example, Ralph's training of WIV employees and Zheng Li and his providing the WIV with the necessary advanced biotechnology to execute GoF work independently did contribute to the creation of SARS-CoV-2. Furthermore, in terms of GoF and DURC policy: "The responsibility for the 'enforcement' of the Guidelines is shared by the NIH Office of Biotechnology Activities, the Recombinant DNA Advisory Committee (RAC), IBCs at individual institutions, and by the principal investigators (PIs) themselves."

Since EHA did not have a BSO, RAC, or IRC, and Peter did not seem to care about any of my concerns related to operational bio risks and bio risk management, it appears at a minimum that he completely and intentionally circumvented the GoF DURC process with creative writing in the specific aims of the NIH-sponsored project and that he lied about GoF and DURC experiments by omission in the proposal.

This is what the Committee on Research Standards and Practices to Prevent the Destructive Application of Biotechnology stated in terms of how the PI, RAC, and BSO are to gain approval for recombinant DNA related work:

> It is also the responsibility of the institution to appoint a Biological Safety Officer if it engages in large-scale research or production activities involving viable organisms containing rDNA molecules. If the institution engages in rDNA research at BL-3 or BL-4 (see below), the officer must be a member of the IBC. The officer's duties include: (1) conducting periodic inspections to ensure laboratory standards are rigorously followed; (2) reporting to the IBC and the institution any significant problems, violations of the Guidelines, and any significant research-related accidents or illnesses; (3) developing emergency plans for handling accidental spills and personnel contamination and investigating laboratory accidents involving rDNA research; (4) providing advice on laboratory security; and (5) providing technical advice to the PI and the IBC on research safety procedures.
>
> Pre-initiation review of experiments by the RAC has been an important part of the oversight mechanism. Pre-initiation approval of experiments by NIH is required only for: (1) experiments that have not been assigned containment levels by the Guidelines; (2) experiments using new host-vector systems, which must be certified by NIH; (3) certain experiments requiring case-by-case approval; and (4) requests for exceptions from Guideline requirements. Prior to the initiation of these experiments the PI must submit a registration document to the IBC containing the following information: the source(s) of DNA; the nature of the inserted DNA sequences; the host(s) and vector(s) to be used; whether an attempt will be made to obtain expression of a foreign gene, and if so, the protein that will be produced; and the containment conditions that will be implemented as specified in the NIH Guidelines.[140]

After reviewing the NIH policy of how GoF and DURC recombinant DNA regulation is supposed to properly function, this begs the question: Was EHA's GoF DURC work ever reviewed by the RAC? I have the answer, but my guess is that it was not reviewed due to Peter's creative writing skills, his propensity to lie, his disregard for US policy and law, and his tendency to take credit for others' work (he did that to me personally).

Here is the answer: In a paper published in July 2015 in the *Journal of Virology* by Zheng Li, Baric, and Daszak, NIH NIAID (Grant #R01AI110964) is listed as the source of funding and it is titled "Bat Severe Acute Respiratory Syndrome-Like Coronavirus WIV1 Encodes an Extra Accessory Protein, ORFX, Involved in Modulation of the Host Immune Response." A vast majority of the paper discusses using recombinant DNA for GoF work on SARS-like coronaviruses (select agents).

Guess what? The NIH proposal referenced, Grant #R01AI110964, is the "Understanding the Risk of Bat Coronavirus Emergence" proposal, and I have the original in my possession. On the original copy, Peter is listed as the PI and the GoF work that was published by the three stooges is completely omitted in the Select Agent Form (page 127), DURC, and GoF. Furthermore, Peter directly violated NIH GoF and DURC policy by pinning any responsibility for biosafety and biosecurity on the WIV in small writing on the Select Agent Form, which is not allowed under official sub-award policy.

Not having a biological safety officer or institutional biosafety committee, lying by omission on the Select Agent Form of the "Understanding the Risk of Bat Coronavirus Emergence" proposal, and passing of the principal investigator's (Daszak) and institutions (EHA) responsibility to oversee are biosafety and biosecurity to a subcontractor (WIV), and these are incredibly severe and egregious Select Agent, GoF, and DURC policy violations and should be enough to result in a ban on federal funding for EHA and Peter. In my opinion, these violations should result in criminal charges against EHA and Dr. Daszak given the six million people that have died globally because of the lies and negligence.

What is completely maddening to me is that the authors of the Fink Report believed that scientists would ethically self-regulate themselves while repeatedly listing laboratory leaks as a serious concern, and then simultaneously downplayed the laboratory leak risks associated with GoF DURC research and development. When creating security related policies, the focus should be a layered approach, which is even stated in the Fink Report; however, they then put policy control and enforcement in the hands of the scientists, university professional staff, and universities.

Many of the infectious disease scientists, virologists, and bacteriologists sit on the biosafety and ethics committees at research institutions, which can then lead to scientists approving their own work or influencing the research institution's biosafety committee.

In my opinion, the policy structure was intentionally written to keep the control of GoF and DURC in the hands of scientists and research institutions. Are these the people that you want securing, protecting, and handling dangerous, pandemic-inducing zoonotic infectious agents to which humans or animals have no immunity?

The ability for scientists and academic institutions to self-regulate is a critical flaw in the DURC GoF policy. There must be third party regulatory oversight outside research institutions that verifies and validates the infectious disease research sponsored by the US government.

The National Science Advisory Board for Biosecurity (NSABB) was established in 2004, and it is a federal advisory committee that addresses issues related to biosecurity and DURC at the request of the US government. The NSABB faced its first major challenge in 2011 over the controversy surrounding the publication of methods for GoF transmissible H5N1 influenza virus.

Most interestingly, the debate about publication shifted to the inherent weaknesses in self-regulation. Others, including my PhD thesis chair, Dr. Michael T. Osterholm, regents' professor at the University of Minnesota, identified the problem that DURC could potentially be self-regulated by scientists, by the government, or by any combination of possibilities. [141] It is highly problematic that an organization like the NSABB cannot enforce its own recommendations.

Additionally, the NSABB has further been plagued with controversy after terminating members of the twenty-five-member committee that were more likely to oppose DURC. [142] The NSABB has been accused of conflicts of interest, and its area of responsibility was reduced in 2014. Are you seeing the pattern here?

The limitations and major weaknesses of laboratory systems are well known—laboratory leaks are bound to occur, especially in the *laissez faire* world of public and private universities. Physical security has always been a problem at universities and academic institutions, so why should they be doing bioweapons work in the first place? If the policy framework to evaluate GoF and DURC was known to have major flaws by allowing non-profit organizations, universities, and scientists to self-regulate the GoF research, which could then be used as bioweapons, then why wasn't the policy corrected?

Clearly, the system is not working domestically, and we cannot trust these people to act ethically.

Domestic GoF and DURC policy must be modified to prevent another pandemic or the annihilation of precious animal or plant species.

Our children and families deserve better.

CHAPTER 16:

The Real COVID Timeline

The timeline of how COVID came into existence extends back as far as all the components related to the disease's emergence and our attempt to control the disease. In this case, the creation of SARS-CoV-2 and the COVID-19 vaccine extends all the way back to 1987 with the invention of the mRNA platform. Well, at least that is what I think.

In 1987, Dr. Robert Malone performed a landmark experiment. He mixed strands of messenger RNA with droplets of fat to create a kind of molecular stew and thus made the discovery which made mRNA vaccines possible. Human cells bathed in this genetic stew absorbed the mRNA and began producing proteins from it. This discovery highlighted the potential of mRNA therapies.[143]

Malone, a graduate student at the Salk Institute for Biological Studies in La Jolla, California, later jotted down some notes, which he signed and dated. If cells could create proteins from mRNA delivered into them, he wrote on January 11, 1988, it might be possible to "treat RNA as a drug." [144]

Later that year, Malone's experiments showed that frog embryos absorbed mRNA2. It was the first time anyone had used fatty droplets to ease mRNA's passage into a living organism.

Those experiments were a steppingstone toward two of the most notorious and profitable vaccines in history: the Moderna and Pfizer mRNA-based COVID-19 vaccines that were given to hundreds of millions of people around the world.

Global sales of mRNA were nearly $50 billion in 2021 alone.[145]

From 1987 to 2008, the mRNA vaccine technology and platform rapidly evolved simultaneously in three key areas and are grouped according to topic, not as they occurred, to enhance clarity:

1. First, the mRNA platform itself was tested as a treatment in rats and as a cancer treatment in mice. In 2005, it was discovered that modified RNA evades immune detection, with the first mRNA clinical trial for infectious diseases occurring in 2013.[146]
2. Second, the lipid-based delivery system advanced with the first report of a four-component lipid nanoparticle, which was used to deliver DNA into cells in 2001. In 2005, a scalable method for developing lipid nanoparticles was invented, and, in 2018, the first drug with lipid nanoparticles was approved by the Food and Drug Administration (FDA).
3. Third, the mRNA platform was combined with the lipid nanoparticle technology to make vaccines, with the first mRNA vaccine developed for influenza and tested in mice in 1993. The first mRNA vaccine with lipid nanoparticles was tested in mice in 2012, and the first clinical trial of mRNA vaccines with lipid nanoparticles was developed and tested for influenza in 2015.

The previous three paragraphs were all related to the mRNA gene therapeutic platform, and this is when the foundation research for the engineering of SARS-CoV-2 enters the story. In 1995, a scientist at the University of North Carolina, Dr. Ralph Baric, made a series of discoveries during interspecies transfer experiments with coronavirus via serial passage (a gain of function technique).[147]

Then, in 2002, Baric's scientists introduced their mouse coronavirus into flasks that held a suspension of monkey cells, human cells, and pig testicle cells.[148] Baric's team had discovered a way to create a full-length infectious clone of the entire mouse-hepatitis genome. They wrote that the "infectious construct" replicated itself just like the real thing.

They also determined how to perform the genetic assembly seamlessly, without any signs of human engineering. The result is that a person would not be able to determine if a virus had been fabricated in a laboratory or grown in nature.

As previously discussed, Baric called this the no-see-um method and he asserted that it had "broad and largely unappreciated molecular biology applications." The method was named after a "very small biting insect that is occasionally found on North Carolina beaches."

These methods, if applied to a virus like SARS-CoV-2, the agent that causes COVID-19, would make it difficult to discern whether an agent was human-engineered.

In 2002, the BioIndustry Initiative's mission was to counter the threat of bioterrorism through targeted transformation of former Soviet biological weapons research and production capacities:

> The U.S. Department of State BioIndustry Initiative (BII) is a nonprolif-eration program authorized in the Defense and Emergency Supplemental Appropriations Act for Fiscal Year 2002 (Public Law 107-117). BII focuses on two objectives: (1) The reconfiguration of former Soviet biological weapons (BW) production facilities, their technology and expertise for peaceful uses; and (2) the engagement of Soviet Biological and Chemical Weapons scientists in collaborative R&D (research and development) projects to accelerate drug and vaccine development for highly infectious diseases.[149]

This was a program run by the US Department of State, which saw numer-ous prominent American biologists and virologists visit former Soviet biowarfare facilities to examine the technology they had developed over the years and see if anything was suitable for patenting and repurposing for civilian use.

Dr. Michael Callahan formerly served as the health director for USAID in Nigeria where he carried out research on deadly pathogens.[150][151]

In 2002, he was hired as the State Department's director for the Bureau of International Security and Nonproliferation to serve as "clinical director for Cooperative Threat Reduction [CTR] programs" at six former Soviet Union Biological Weapons facilities as part of the BII program. He was officially tasked with carrying out the stated goals of the mission, which entailed the "reconfiguration of former biological weapons production facil-ities" in the former Soviet Union and the acceleration of "drug and vaccine production." Callahan would be put in charge of gain of function programs and for viral biological agents at these facilities.[152][153]

From 2002 through 2003, there was a large SARS outbreak, which frightened the globe and generated much interest in these new types of coronaviruses and the opportunities to conduct GoF research involving them. In fact, a patent war broke out over the intellectual property rights for the genetic material of the agent SARS-CoV-1.[154]

In 2005, Ken Alibek, a Kazakh American microbiologist and biological warfare administrative management expert, testified before congress alongside Dr. Callahan. Ken Alibek rose rapidly in the ranks of the Soviet Army to become the first deputy director of Biopreparat, with a rank of colonel, during which time he oversaw a vast program of biowarfare facilities. This testimony dramatically increased DARPA's budget and the Department of Defense think tanks budgets increased by hundreds of millions of dollars. Callahan stated during his testimony:

> The dark science of biological weapon design and manufacture parallels that of the health sciences and the cross mixed disciplines of modern technology. Potential advances in biological weapon lethality will in part be the byproduct of peaceful scientific progress. So, until the time when there are no more terrorists, the U.S. Government and the American people will depend on the scientific leaders of their field to identify any potential dark side aspect to every achievement…"[155]

In 2005, Senators Barack Obama and Dick Lugar oversaw the establishment of jointly operated laboratories and epidemiological monitoring stations in Ukraine, with Victor Yuschenko's approval, and with the participation of the United States Department of Defense.[156] Senator Lugar was one of the longest serving senators in history and is responsible for the Cooperative Threat Reduction Program as part of the Nunn-Lugar Act. The Nunn-Lugar Act was a great program, in my opinion and based on data, to reduce nuclear threats abroad at the end of the Cold War, in which Sandia National Laboratories was heavily engaged.[157] These types of programs were instrumental in reducing chemical, biological, radiological, and nuclear threats in the former Soviet Union.[158] In many cases, these diplomatic relationships between scientists and engineers in the United States and abroad have prevented the proliferation or dissemination of weapons of mass destruction.[159] In my opinion, as the size and complexity of these well-intended and effective programs increased, they became too difficult for government officials to effectively manage. As it appears, at least in the case of the laboratory creation of SARS-CoV-2, the lack of effective program management may have been abused by some members of the US government to outsource GoF research during the domestic pause on GoF research.

In 2006, the Accelerated Manufacture of Pharmaceuticals (AMP) program was created by Dr. Callahan, barely a year after he first came on board as DARPA's portfolio manager.[160] Its purpose was to find technologies that could "radically accelerate the manufacturing of protein vaccines and protein-based therapeutics," with the goal of "revolutionizing protein therapeutics and vaccine manufacture" through the private sector. In my opinion, there is a real need for rapidly developed therapeutics and vaccines due to the significant amount of time it takes to run randomized controlled trials (RCTs), and the goal of this program was to address this need; however, they should be limited in scale and not be applied to the entire population as in the case of the mRNA COVID-19 "vaccines."

In September 2008, the NIH awarded Dr. Peter Daszak of EcoHealth Alliance $535,136 to study emerging zoonoses. The proposal factually argued that zoonoses are a major threat to health globally, causing tens of thousands of deaths each year in the United States, and that many zoonoses have recently emerged from bats (e.g., SARS, Ebola, Nipah).[161]

The proposal claimed that the research would provide a way to predict the regions where the next new emerging zoonoses from bats was most likely to occur and proposed targeted surveillance of these animals using state-of-the-art molecular techniques in those regions. The proposal stated that it would characterize new viruses, study the pathogenesis of these new viruses, and collect a bank of identified bat viruses that have not yet emerged in the human population.

The proposal further, and incorrectly, argued that the work would be predictive of emerging diseases and claimed that the approaches would proactively combat the most high-profile group of emerging pathogens.

Both are very bold scientific claims.[162]

Just as Callahan was soliciting proposals and handing millions in DARPA funding to private companies, the agency was entering into a cooperative agreement (HR0011-07-2-0003) with the University of Pittsburgh Medical Center (UPMC) to investigate the challenges associated with the endeavor. The program's mandate dovetailed with concurrent efforts (2006-2009) to fundamentally transform the US government's approach to vaccine manufacture and Medical Countermeasures (MCMs) to protect military and civilian populations against a chemical, biological, radiological, and nuclear (CBRN) attack and naturally occurring outbreaks of emerging infectious diseases.[163]

In 2009, the seminal 180-page report that resulted from the two-year deep dive into US government procurement and manufacturing methods for MCMs was published by DARPA.

It was titled "Ensuring Biologics Advanced Development and Manufacturing Capability for the US Government: A Summary of Key

Findings and Conclusions" and was led by Tara O'Toole and Thomas Inglesby. They were both key individuals in the Dark Winter exercise, a simulated tabletop exercise of an infectious disease pandemic in the United States, and were perennial participants during pandemic response policy and legislative changes, including medical counter measure development.

The central question that this cooperative effort between DARPA and the UPMC wanted to answer was how to incentivize the private sector to manufacture products that only had one buyer, the US government. To this end, the researchers probed different areas such as barriers to entry, cost analysis, and several types of manufacturing options.

They included one case study to demonstrate what they believed would be the most effective strategy to follow.

That case study looked at a company headquartered in Rockville, Maryland, called Novavax, which had recently received a $1.6 billion grant (the largest so far) from Trump's Operation Warp Speed to manufacture a COVID-19 vaccine.[164]

The analysis determined that liability immunity was required for vaccine manufacturers, otherwise vaccine manufacturers would not be willing to manufacture vaccines due to the high liability and risk associated with a shortened vaccine safety and evaluation period.

This policy recommendation included liability immunity for vaccine manufacturers and was eventually adopted into the Public Readiness and Emergency Preparedness Act for Medical Countermeasures Against COVID–19 in 2021. [165]

In 2009, Michael Callahan's old employer USAID launched PREDICT, an early warning system for new and emerging diseases, in twenty countries.

In 2009, Dennis Carroll, a former USAID director of the emerging threats division who had led the United States' response to Avian influenza (H5N1) in 2005, would go on to create PREDICT. USAID partnered with a non-profit called EcoHealth Alliance to carry out its nine-year effort to catalog hundreds of thousands of biological samples, "including over 10,000 bats." UC Davis was the prime contractor, and Metabiota and EHA received subawards on the prime contract (a.k.a. the group was known as the PREDICT partners).

Also in 2009, the Wuhan Institute of Virology (WIV) in Wuhan, China began collaborating with EcoHealth Alliance on the USAID Emerging Pandemic Threat program on a project titled PREDICT. The PREDICT partners were USAID, UC Davis, Wildlife Conservation Society, EcoHealth Alliance, Metabiota, and the Smithsonian Institute and PREDICT and

USAID are listed in the letters of support from collaborators in China in the "Understanding the Risk of Bat Coronavirus Emergence" proposal.[166] [167]

By identifying unknown viruses before they spilled over into humans—to "find them before they find us," as WIV virologist Shi Zhengli put it—researchers hoped to establish an early-warning system. PREDICT worked in dozens of countries, but the WIV was one of its linchpins, and Shi Zhengli became famous as China's "Bat Woman."[168]

PREDICT was a project within USAID's Emerging Pandemic Threats (EPT) program and was initiated in 2009 to strengthen global capacity for detection and discovery of zoonotic viruses with pandemic potential. Those included coronaviruses, the family to which SARS and MERS belong; paramyxoviruses, like Nipah virus; influenza viruses; and filoviruses, like the Ebola virus.

PREDICT claimed that it made significant contributions to strengthening global surveillance and laboratory diagnostic capabilities for new and known viruses with partners in twenty countries which expanded over time to roughly thirty countries. PREDICT claimed to be building platforms for disease surveillance and for identifying and monitoring pathogens that can be shared between animals and people.

Using the "One Health" approach, which is public health theory which includes all aspects of the environment, species, and behavior, PREDICT investigated the behaviors, practices, and ecological and biological factors driving disease emergence, transmission, and spread. Through these efforts, PREDICT claimed that it would improve global disease recognition and develop strategies and policy recommendations to minimize pandemic risk.[169]

The stated goal of this program was to conduct surveillance of pathogens that could be a zoonotic spillover risk.

This seemed reasonable on the surface.

Some of the worst diseases in human history started in wildlife and made the jump to people. However, the way spillover risk was "predicted" in the PREDICT program, was by merely sampling zoonotic animal viruses and analyzing the agents in computational models to determine the likelihood that they would evolve to have "pandemic potential." Then EHA used the relationships formed with China and the samples collected during PREDICT to execute the gain of function (GoF) work described in "Understanding the Risk of Bat Coronavirus Emergence" proposal to conduct gain of function research on them to make them capable of infecting human cells, which is a process that is indistinguishable from bioweapon research.

Therefore, they refer to it as dual use research of concern (DURC).

In 2009, the relationship between the WIV and the American bio-defense establishment was advanced by EHA policy advisor Dr. David R. Franz, former commander at the US bioweapons lab at Fort Detrick (USAMRID).[170] [171] Franz was chief inspector on the three UN Special Commission biological warfare inspection tours in Iraq, which included a young Dr. Robert Kadlec.

Robert Kadlec served as a career Air Force officer and physician and later served as the assistant secretary of Health and Human Services, Administration for Strategic Preparedness and Response (ASPR).[172]

Franz was a member of the team on the ground, and advised Robert Kadlec as a member of HHS' National Science Advisory Board for Biosecurity (NSABB), the organization that reviews the suitability and ethics of GoF and DURC.[173] Kadlec was an ad-hoc voting member of the NSABB.

Moderna files several patent applications on or about 2013, and the furin cleavage site is found referenced later in SARS-CoV-2 patent applications filed by Moderna in 2016. The probability of the patented man-made sequence appearing in nature, via natural evolution and natural selection, is one in the billions and is statistically impossible. [174] [175] [176] [177] [178] [179] [180] [181] This strongly suggests that the infectious agent SARS-CoV-2 and the COVID-19 mRNA vaccine were co-developed. In my humble opinion, it is impossible that they were not co-developed.

In September 2014, I was hired as the senior scientist of Data & Technology at EcoHealth Alliance. I quickly learned that I had to turn around a failing department that had personnel issues, a lack of direction, and was not performing well.

In late 2014, I was asked to prepare a report for the Intelligence Advanced Research Projects Activity, Office of the Directorate of National Intelligence, (IARPA). I later learned, upon promotion to Associate Vice President, while attending weekly finance updates that EcoHealth Alliance did not receive any funding from this agency (IARPA), as far as I am aware.

In 2014, gain of function research was a highly contentious topic in my scientific area of expertise. Those who were for it made the argument that if you can identify a high-risk pathogen, and then engineer the pathogen in the laboratory to increase its transmissibility, infectivity, pathogenicity, or virulence, then you can develop medical countermeasures to prevent the spread of disease, if an outbreak of a naturally evolving agent were to occur. I believe this logic to be inherently flawed because it is naïve to think that humans can modify or engineer a naturally occurring pathogen that would evolve similar to the way infectious agents naturally evolve. Typically, gain of function research (via selection of rare traits or genetic manipulation or engineering of the agent) undergoes thousands of years of unnatural evolution (decided by humans, not by nature) in a laboratory in a matter of days, weeks, or months. This is akin to predicting the future, with the likelihood of success decreasing in every timestep.

Simultaneously, in the late fall of 2014, I received an electronic copy of a proposal in PDF format either in preparation to be submitted to the National Institutes of Health (NIH) or being edited presumably for negotiation or resubmission, National Institute of Allergy and Infectious Diseases (NIAID), managed by Dr. Anthony Fauci, titled "Understanding the Risk of Bat Coronavirus Emergence."

I was asked to review (provide edits, comments, and feedback) the proposal without any rationale or justification.

I reviewed the proposal that was submitted to NIH, which detailed the gain of function virology work that was being conducted to create the agent known as SARS-CoV-2, which causes the disease now known as COVID-19.

I initially learned from reviewing the proposal, that EcoHealth Alliance was working with the Wuhan Institute of Virology and with Dr. Ralph Baric at the University of North Carolina to conduct SARS-CoV-2 GoF research.

In the proposal, the coinvestigators in the United States and China stated that they were working on the GoF work before the receipt of the NIH NIAID funding which was supported by USAID.

The proposal clearly stated that the gain of function work on SARS-CoV-2 was already underway in China, prior to October 2014, at the WIV, with the support of USAID in collaboration with EcoHealth Alliance and EcoHealth Alliance's partners and sponsors.

In 2014, I made Dr. Peter Daszak aware of the lack of a Biological Security Officer (BSO) and Institutional Biosafety Committee (IBC) at EcoHealth Alliance in reference to the Select Agent Form in the "Understanding the risk of Bat Coronavirus Emergence" proposal, in accordance with NIH requirements.

From 2014 to 2016, I also witnessed firsthand presentations by Dr. Shi Zhengli (WIV) and Dr. Ralph Baric (UNC) at EcoHealth Alliance related to their Gain of Function work managed and supported by EcoHealth Alliance.

From 2014 to 2016, I also witnessed firsthand presentations by the executive team at EcoHealth Alliance related to the gain of function work conducted and managed at EHA with Dr. Daszak listed as the principal investigator on the PREDICT subaward and NIH awards.

EcoHealth Alliance developed SARS-COV2 and was responsible for the development of the agent SARS-COV2 during my employment at the organization (from 2014 to 2016).

In 2015 and 2016, during an executive meeting, I informed the EcoHealth Alliance executive team that I believed there were biosafety and biosecurity risks in contract laboratories. Specifically, I was concerned that EcoHealth Alliance did not have enough visibility or firsthand knowledge of what was happening at foreign laboratories contracted and managed by EcoHealth Alliance. During this meeting I discussed biorisk management with the team due to my observations and concerns. Dr. Daszak refused to mitigate the risks without any objection or discussion from the other executives. In my opinion Daszak was dismissive of my concerns. He did not seem concerned about EcoHealth's lack of oversight which I felt was strange because it is typically the CEO's duty to protect the organization from organizational threats and risks. After raising my concern, I accepted Peter's position that he felt that our organization's control measures were adequate (they were not).[182] [183] [184]

In 2015, at the age of thirty-two, I successfully turned around the department and succeeded at receiving a $4.6 million dollar contract from the Defense Threat Reduction Agency (DTRA). I improved the relationship with US government project sponsors and my work was being published in the mainstream media.

On or around June 2015, I was promoted to vice president. After being promoted to vice president, I was exposed and participated in more aspects of the organization, as would be expected from an Executive Officer at any organization.

In 2015, before my promotion, I made the assessment that the PREDICT project could not meet the claims of predicting emerging infectious diseases as the sample sizes are too small, sampling is infrequent, the project design is incorrect, and the methods applied to the collected data are incorrect. I believe and conclude that the project is mostly a global fishing expedition for coronaviruses. Several other experts outside EHA make the same conclusions about PREDICT. After promotion, I began attending weekly financial meetings where I learned that the organization was tight on cash, depended heavily on government contract salary overhead to remain solvent, and that the organization was not involved in traditional conservation work as classically defined. This was upsetting as this was one of the main reasons that I wanted to join the organization (being a conservationist and naturalist). I was very displeased to learn in an executive meeting that no money was spent on conservation. Additionally, I determined that EHA was more closely aligned with biodefense, which was a personal advantage for me personally due to my experience in the space and understanding the customers.

Upon being promoted to assistant vice president in June of 2015, I made requests to Peter Daszak during my promotion, one of which was that I be added to the PREDICT program, which I was.

Also, from observing security risks to the business, I requested to assume an additional role of chief security officer/chief technology officer, and I identified numerous security concerns related to cyber, information, physical, and biological security.

Daszak did not see the operational security risks or did not value dedicating resources to mitigate the risks and rejected my request to lead security and safety at EHA.

After being promoted to vice president, I commented on several concerns I had related to protecting the organization. My primary concern was related to my opposition to gain of function research. None of the other executives voiced any opposition to gain of function research being conducted at EcoHealth Alliance, and Dr. Daszak was heavily supportive of the work. Drs. Johnathan Epstein and Kevin Olival were supportive of the work and were key contributors to the gain of function work in the SARS-COV2 proposal funded by USAID and NIH, and executed by EcoHealth Alliance, the WIV, and UNC. My opposition to gain of function research stemmed from my PhD studies taught by my Committee Chair, Dr. Michael T. Osterholm, who would become President Joe Biden's COVID advisor.[185]

In November 2015, a scientifically peer reviewed, and referenced article was published by collaborators from the Wuhan Institute of Virology, the University of North Carolina Chapel Hill (UNC), the Food and Drug Administration, Harvard Medical School, and the Bellinzona Institute of Microbiology. The peer reviewed article was titled "A SARS-like Cluster of Circulating Bat Coronaviruses Shows Potential for Human Emergence" in the journal Nature Medicine.[186] The authors initially omitted the funding source from the USAID - EPT - PREDICT program, of which I was a co-investigator and country coordinator while employed by EHA. The USAID - EPT - PREDICT funding cited in the article was used to develop a relationship between Drs. Ralph Baric (UNC) and Shi Zhengli of the Wuhan Institute of Virology at EcoHealth Alliance, which was orchestrated by Dr. Peter Daszak. Additionally, the USAID- EPT - PREDICT funding used in this peer reviewed paper was used to collect biological samples from bats globally. Then, the collaborators analyzed the collected samples to extract SARS-like coronaviruses, and select or engineer genetic features within the viruses, collected with USAID - EPT - PREDICT funding, to create hybrid chimeric viruses. Chimeric viruses are defined as combining the genetic material from two or more distinct viruses. The process of developing SARS-COV2 was also described in detail in the proposal submitted to, and ultimately funded by, the National Institutes of Health (HHS NIH), The National Institute of Allergy and Infectious Diseases (NIAID), by EcoHealth Alliance with the WIV and UNC listed as collaborators. It is my attestation, that the creation of these SARS-like chimeric viruses described in this article include SARS-COV2. Lastly, the engineered SARS-COV2 was then used to test SARS vaccines and monoclonal antibody therapeutics against the disease in mice.

Dr. Peter Daszak approached me in late 2015 and stated that somebody from the Central Intelligence Agency (CIA) approached him and stated that they were interested in the places we were working, the people we were working with, and the data we were collecting. Peter then proceeded to ask me for my advice, and specifically asked whether we should work with them. I was shocked that Peter asked me this and was excited for the opportunity. I stated to Peter that "It never hurts to talk with them. There could be money in it." Peter then later confirmed over the next two months, between our weekly meetings, that the relationship with them was proceeding.

In 2015 and 2016, I also observed that EcoHealth Alliance was engaged in irregular financial transactions regarding U.S. Government grants. Specifically, I believe there was timecard fraud and observed what appeared to be double dipping on contracts, or material support, between government organizations and private donors (e.g., Skoll Foundation, Google Foundation, Rockefeller Foundation, & Wellcome Trust), or both.

In 2016, I confronted Dr. Peter Daszak, Harvey Kasdan (CFO, deceased), and Dr. Aleksei Chmura about the financial fraud when I was upset and arguing for pay raises in my department, company-wide salary increases, and for myself. Shortly thereafter (one to two days), CFO Harvey

Kasdan passed away from a heart attack. I am not insinuating foul play, but I believe the stress was too much for him in his physical condition.

In 2015 and 2016, while attending board meetings and in communications directly with board members, that Dr. Peter Daszak had a pattern of over-simplifying and lying by omission to our stakeholders (including the board of directors). For example, while EcoHealth Alliance positioned itself as a conservation organization, no substantial conservation work, as traditionally defined, was occurring at the organization.

In 2016, I draw strong conclusions about the PREDICT program. The USAID Predict program was a global hunt for viruses predicated upon the promise of predicting and preventing pandemics. I believe that the data limitations and methods for collecting and analyzing that data make this goal impossible to achieve. I further believe that this program is more strongly aligned with collecting the biological samples to conduct gain of function viral work, or intelligence collection, than prediction and prevention of pandemics.

During my tenure at EcoHealth Alliance (2014–2016), UC Davis was the prime contractor on the PREDICT award, and EcoHealth Alliance and Metabiota were co-leading sub-contractors. The nature of the relationship between the entities was competitive yet cordial. It was my observation that the project was heavily micromanaged by USAID personnel, US embassy or consulate staff, and employees of the State Department. I recall that the project was not like any other federally funded project that I had been a part of. The micromanagement was unlike anything that I have ever experienced. At the time, I felt like the project seemed more like intelligence collection than scientific research and development. In 2015, during the PREDICT project planning meetings with the executive team, the executive team openly discussed the merits and safety of gain of function work. I wanted to make it clear that I was opposed to gain of function work due to the numerous accidental laboratory leaks that have occurred throughout history, and due to the flawed logic of pathogen gain of function research.

The flawed logic is that human laboratory engineered viruses evolve similarly to naturally occurring viruses; however, there is no scientific evidence to support this logic. Therefore, scientists cannot produce a vaccine based on gain of function work, to match what would evolve and emerge naturally in nature.

I observed at EcoHealth Alliance that the organization commonly worked ahead of the receipt of federal funding. Meaning that EHA would execute research in contract and grant proposals before they were awarded or funded by federal agencies. This is a common practice throughout academia to obtain the necessary preliminary data to obtain funding from the highly competitive grant process.

EcoHealth Alliance and foreign laboratories did not have the adequate control measures in place for ensuring proper biosafety, biosecurity, and risk management, ultimately resulting in the lab leak at the Wuhan Institute of Virology. I raised these concerns at an executive project planning meeting, where Daszak quickly dismissed my concerns. This information was later validated by cables between the US Consulate in China and the State Department.

Although, the problems of managing biosafety and biosecurity risks from abroad was not limited to the partnership with China.

During my tenure at EHA, I was asked to review and contribute to an investment "pitch deck" (i.e., a PowerPoint presentation used in venture capital presentations) that was presented to an organization called In-Q-Tel. In the pitch deck, we proposed an extension of the USAID global disease surveillance work, SARS-CoV-2 gain of function and humanized mice research conducted by Drs. Baric and Zhengli, and my work from my department developing advanced biosurveillance technologies and platforms. This work was presented to In-Q-Tel (which can be verified in In-Q-Tel's Quarterly Report). I do not know the outcome of that meeting as it was not communicated to me by Dr. Daszak.

In-Q-Tel (IQT) is a Department of Defense and Central Intelligence Agency venture capital firm. IQT invests in companies that make technology that is of national security interest.

In the IQT pitch deck, the work Daszak discussed was the humanized mice and SARS-CoV-2 gain of function work being conducted at Ralph Baric's laboratory at the University of North Carolina, and at the Wuhan Institute of Virology.

Additionally, Daszak requested that I write a report to the Intelligence Advanced Research Project Agency (IARPA). However, I was not aware of receiving any funding from IARPA, which made writing and submitting a report to IARPA strange.[187]

In 2015, during the PREDICT project planning meetings with the executive team, the executive team openly discussed ongoing research in China at the Wuhan Institute of Virology.

In attempting to protect EcoHealth Alliance, I was opposed to the work in China for several reasons. I was concerned with providing the training to the Chinese Communist Party in advanced biotechnology methods, which could be used to make biological weapons, and the risk of intellectual property theft from EcoHealth Alliance and the United States.

My concerns were acknowledged and dismissed by Daszak.

In 2015, I made the assessment that the PREDICT project could not meet the claims of predicting emerging infectious diseases, as the sample sizes were too small, sampling was infrequent, the project design was incorrect, and the methods applied to the collected data were incorrect.

I believed and concluded that the project was mostly a global fishing expedition for coronaviruses. Several other experts outside EHA made the same conclusions about PREDICT.

In March 2016, a paper was published by Dr. Ralph Baric, an EcoHealth Alliance gain of function collaborator working at UNC, in PNAS titled "SARS-like WIV1-CoV Poised for Human Emergence." In the article, the authors of the paper describe in detail how they used, designed, and constructed full-length and chimeric viruses to determine if they would replicate in human airway cultures.[188] This specific paper is relevant because it compares and documents the effectiveness of different variations of coronavirus spike proteins at infecting human cells specifically by binding to ACE2 receptor, which was a critical and necessary step to design and engineer the SARS-CoV-2 virus. While employed at EcoHealth Alliance, I met both Dr. Shi Zhengli and Dr. Ralph Baric, when they presented their work on the design and engineering of SARS-CoV-2 (coronavirus gain of function research), and the use of highly specialized humanized mice models, which were necessary to successfully build SARS-CoV-2. These facts are supported by numerous recorded presentations by Dr. Peter Daszak and Dr. Ralph Baric from 2015–2019, some of which I personally attended while employed at EcoHealth Alliance. Additionally, the specific gain of function work described in this paper was presented by Dr. Peter Daszak to In-Q-Tel, a DoD and CIA venture capital firm. In the slides presented to In-Q-Tel, which I personally helped create at EcoHealth, the use of USAID- EPT-PREDICT funding to collect coronavirus samples from bats globally is described, where they are then analyzed to identify their most dangerous features to humans, and recombined to make new coronaviruses like SARS-CoV-2. Then, these viruses are tested on humanized mice to validate lethality and transmissibility. EcoHealth Alliance then used Dr. Baric's work for testing experimental vaccines, treatments, and therapeutics against the newly engineered SARS-CoV-2 strain to determine which countermeasures would be the most effective at mitigating the disease in humanized mice.

In 2015 and 2016, Peter Daszak publicly discussed the gain of function work at UNC and WIV at several conferences and meetings. During the discussions, he claimed that the gain of function work was necessary to produce vaccines and other MCMs to counter laboratory engineered agents like SARS-CoV-2.

For the evaluation period of 2015, I received a near perfect performance evaluation from Daszak in late 2015 and early 2016.

Yet I finally decided to leave EcoHealth in early January 2016 due to the large number of ethical concerns with the scientific work and EHA as whole, and I began to interview for new positions.

In 2016, I witnessed that EcoHealth Alliance had what appeared to be a very tight cash flow and that EHA engaged in what I believed to be fraud against the US government (timecard fraud contract, reimbursement fraud, and potentially double dipping).

After I identified and learned about these serious issues, I brought them to the attention of Peter Daszak, Dr. Aleksei Chamura, and CFO Harvey Kasdan.

After raising these issues at the meeting, Harvey Kasdan went home from work, had a heart attack, and died.

In 2016, Daszak, a regular advisor to WHO on pathogen prioritization for Research & Development (R&D), Dennis Carroll, the creator of PREDICT, and Jonna Mazet, former global director for USAID's PREDICT, all formed together the Global Virome Project; a "10-year collaborative scientific initiative to discover unknown zoonotic viral threats and stop future pandemics".[189] [190]

Dr. Jonna Mazet, the overall Principal Investigator of PREDICT, was also co-director of UC Davis' One Health program, which recruited Dr. Wacharapluesadee and her team in Thailand to conduct a multi-year research project on bats.[191] They were joined by Edward Rubin of Metabiota, a recipient of DARPA's project PROPHECY funds (a program led by Dr. Callahan to predict infectious diseases).[192] Notably, they received an $18.4 million DTRA contract award for scientific research and consulting work in Ukraine and the Lugar Center in the Republic of Georgia.[193] [194] Metabiota was accused by the Viral Hemorrhagic Fever Consortium in 2014 of violating their contract and engaging in dangerous blood culturing work at a lab in Africa, as well as misdiagnosing patients.[195] [196]

In 2017, during a visit to the WIV as part of the "Second China-U.S. Workshop on the Challenges of Emerging Infections, Laboratory Safety and Global Health Security,"[197] EHA policy advisor David Franz outlined "possible joint project ideas," which included carrying out joint "table top exercises" or simulations of outbreaks (e.g. exercises similar to Dark Winter), decision-making related to gain of function research, and "overcoming barriers to sharing strain collections and transport of pathogens." Sharing of genetic material was a key component of the design and engineering of SARS-CoV-2.

The last point would play a crucial role in the narrative about the origins of the SARS-CoV2, which has been claimed to be the WIV itself.

EcoHealth's Executive Vice President William Karesh links directly back to the very top of the US biodefense establishment as a member of ASPR Robert Kadlec's original Blue-Ribbon Panel on Biodefense, along with Hudson Institute senior fellows Tevi Troy, Jonah Alexander, and Scooter Libby, whose pivotal roles have been detailed in the Engineering Contagion series by journalist Whitney Webb.[198]

EHA is listed as a partner of the WIV on archived pages of its website and was mentioned as one of the institute's "strategic partners" by the WIV's Deputy Director General Prof Yanyi Wang in remarks during the visit of an official US delegation to the institute in 2018.[199] [200]

During August 2019, the first cases of COVID-19 have emerged in China and are spreading globally. From August 2019 to January 1, 2020, several different strains were identified from different genetic lineages. [201] [202] [203] [204] [205] [206] [207] [208]

In September or early October 2019, I was contacted by Dr. Amy Jenkins, who attempted to recruit me to be a program manager for emerging infectious disease work at the Defense Advanced Research Projects Agency (DARPA). I first met Dr. Amy Jenkins as a PhD student and paid research fellow at a Department of Homeland Security Center of Excellence at the University of Minnesota in 2014, and I was told that she was working with a program named ARGUS-BIO.[209] The position at DARPA was presented to me as if it was mine if I wanted it, and I was told that it would need Top Secret Security clearance with a polygraph. I felt that the recruitment effort was quite strange as I had not worked full-time and directly in the national security space since 2014 at Sandia National Laboratories and I had no clue how Dr. Jenkins obtained my new personal cell phone number.

Coincidentally, this is when epidemiological evidence indicates that the first cases of COVID-19 likely emerged.

The two events may not be related; however, it is my belief that people working within the US government potentially identified me as a risk to knowing firsthand that the SARS-CoV-2 disease emergence event was a consequence of the US government's sponsorship of the genetic engineering of SARS-CoV-2 domestically and abroad. If I would have accepted the position, then I suspect that DARPA would have disclosed restricted information to me, which would have consequently prevented me from discussing any of this information publicly, like I have been and am doing now. The recruitment effort itself was highly suspect as it seemed as if DARPA was completely circumventing the US government recruitment process for one of the most prestigious scientific positions in the world.

I declined the offer to apply and interview for the position. At the time, I thought it was very strange that I was being recruited for this highly competitive position as I intentionally decided to leave the classified infectious disease career field for personal reasons.

On December 12, 2019, a material transfer agreement from NIH NIAID and Moderna to UNC Chapel Hill and Ralph Baric for mRNA coronavirus vaccine candidates to be developed and jointly owned by NIAID and Moderna was approved.[210] [211] This suggests that Moderna and NIH were already in possession of SARS-CoV-2 no later than December 12, 2019.

This was before the SARS-CoV-2 pandemic officially began, according to US and Chinese government officials.[212]

On May 5, 2020, a peer-reviewed scientific paper was published citing Ralph Baric's work from UNC:

> SARS-CoV-2 is a circulating vaccine-derived-coronavirus (cVDCV) borne from work originally done at the University of North Carolina [Ralph Baric's lab], the only institution on earth that's been attempting to design a live-attenuated vaccine for SARS, where they also pioneered engineering the sort of SARS-like chimeric coronaviruses that would be needed as templates for attenuation, and did their best to ignore or circumvent restrictions on gain of function research, and obfuscation that's still ongoing as they refuse to disclose genomic details relating to lab accidents that occurred during the above publicly-funded research.[213] [214]

In October 2020, HHS Secretary Francis Collins and Anthony Fauci criminally conspired to smear "fringe epidemiologists" like myself who did not agree with their narrative. I was an early signatory of the Great Barrington Declaration and vocal critic of many of the COVID-19 response policies.[215] [216]

On December 16, 2013, they applied for four patents with US9149506B2, US9216205B2, US9255129B2, and US9301993B2.[217] [218] [219] [220] Moderna had developed the nineteen-nucleotide gene sequence containing the furin cleavage site which gave SARS-CoV-2 its infectivity to humans by patented gain of function research as early as 2013, six years before the Wuhan outbreak took place.

It was later reported that the first cases of COVID-19 were detected at the Military World Games in Wuhan, China just before this time.

In 2020, mRNA COVID-19 vaccines received emergency authorization from the FDA under artificial conditions due to the US government's suppression of other effective treatments.

According to FDA regulation, an emergency use authorization is only warranted if "there are no adequate, approved, and available alternatives."

On February 21, 2021, there was a complete match found between Moderna's 2016 patent application and the genetic sequence of SARS-CoV-2 circulating in humans. This would have been virtually impossible unless the Moderna vaccine from 2016 and the pathogen that emerged in China were co-developed. In March 2020, Health and Human Services Secretary Francis Collins issued a PREP Act Declaration covering COVID-19 tests, drugs, and vaccines, which provided liability protections to manufacturers, distributors, states, localities, licensed healthcare professionals, and others identified by the secretary who administer COVID-19 countermeasures.[221] [222]

The Declaration has been amended several times to expand liability protections, including prior amendments to cover licensed health-care professionals who cross state borders and federal response teams.[223]

In the spring of 2020, I contacted US Army Surgeon General Scott Dingle to re-enter and receive a direct commission into the Army to assist with the COVID-19 pandemic response at the rank of Lt. Colonel or Major. General Dingle referred me to his deputy commander, and I submitted a completed direct commission packet to the US Army to assist with COVID-19 response as an expert in emerging infectious diseases. My packet was not processed by United States Army Recruiting Command for unknown reasons. I followed up with Gen. Dingle and his deputy in the fall of 2020 and met with the deputy commander's team to discuss what occurred.

I intended to communicate what I knew about the origin of COVID with US army leadership once I re-entered the service. Gen. Dingle's staff stated that my direct commission packet was very strong and that they would follow-up with me. No response from the deputy was received. From March 2020 to early October 2021, I worked with several prominent journalists, including Miranda Devine from the *New York Post*, to obtain the DARPA DEFUSE proposal which was found and published by former Marine Corps weapons of mass destruction expert Charles Rixey from DRASTIC.

Additionally, I was working with Jan Jekielek, editor-in-chief from the *Epoch Times*, Dr. Brett Weinstein, a theoretical biologist who is often labeled as an independent critical thinker, and Alex Berenson, an independent journalist formerly of the *New York Times*.

At no time was my name provided as the source of the information regarding the CIA's involvement with Daszak or EHA in the publications.

From September 2021 to October 2021, I sent copies of all the files that I'd saved from my employment at EcoHealth Alliance to journalists Alex Berenson, Jan Jekielek, Miranda Devine, Dr. Brett Weinstein, and Katherine Eban, a journalist employed by *Vanity Fair*. The files contained seven gigabytes of video files and one gigabyte of documents and were later reported on by Katherine Eban in *Vanity Fair*.

These discussions resulted in publications indicating that Dr. Peter Daszak, president of EcoHealth Alliance, was working with the CIA, and that the biological agent commonly known as COVID-19 (SARS-CoV-2) had been in development at EcoHealth Alliance since 2012, and other evidence suggested that SARS-CoV-2 began earlier than 2012.

The development of SARS-CoV-2 included several prominent US-based scientists and US academic institutions that received funding from numerous federal government agencies and private non-governmental organizations to complete the gain of function work on SARS-CoV-2.

This work was completed domestically and abroad in partnership with several countries for sample collection, analysis, and laboratory work including gain of function work, which was primarily performed at Columbia University, the University of North Carolina, and at the Wuhan Institute of Virology in China.

This has been corroborated by other experts under oath (Dr. Peter McCollough and Dr. Richard Fleming).[224] [225] [226]

On August 8, 2021, US Marine Corp. Major Joseph Murphy's report to the Department of Defense inspector general, said SARS-CoV-2 is "a synthetic spike protein chimera engineered to attach to human ACE-2 receptors and inserted into a recombinant bat SARSr-CoV backbone."[227]

On November 11, 2021, the US Department of Health and Human Services added "SARS-CoV/SARS-CoV-2 chimeric viruses resulting from any deliberate manipulation of SARS-CoV-2 to incorporate nucleic acids coding for SARS-CoV virulence factors" to the list of "biological agents and toxins listed in this section that have the potential to pose a severe threat to public health and safety."[228]

On February 21, 2021, there is a complete match found between Moderna's 2016 patent application and the genetic sequence of SARS-COV2 circulating in humans. This is virtually impossible unless the Moderna vaccine from 2016 and the pathogen that emerges in China were co-developed.

On March 14, 2022, documents were published confirming Moderna created or received genetic material discovered in the SARS-CoV-2 virus. Moderna applied for a patent not only on the reverse compliment of the twelve-nucleotide furin cleavage site in COVID-19 but on the nineteen-nucleotide sequence containing it as previously described. Furthermore, they did not merely apply for a single patent on February 4, 2016, with US9587003B2, as reported in the *Daily Mail*.[229]

These documents were published confirming Moderna created or received genetic material discovered in the SARS-CoV-2 virus. Moderna applied for a patent not only on the reverse complement of the twelve-nucleotide furin cleavage site in COVID-19 but on the nineteen-nucleotide sequence containing it as described below:

> A peculiar feature of the nucleotide sequence encoding the PRRA furin cleavage site in the SARS-CoV-2 S protein is its two consecutive CGG codons. This arginine codon is rare in coronaviruses: relative synonymous codon usage (RSCU) of CGG in pangolin CoV is 0, in bat CoV 0.08, in SARS-CoV 0.19, in MERS-CoV 0.25, and in SARS-CoV-2 0.299. A BLAST search for the 12-nucleotide insertion led us to a 100% reverse match in a proprietary sequence (SEQ ID11652, nt 2751-2733) found in the US patent 9,587,003 filed on Feb. 4, 2016.[230]

In my opinion, many of the people listed in this chapter behaved like a pharmaceutical pseudoscience mafia entrenched in the halls of the medical military industrial complex.

CHAPTER 17

The Truth about Wuhan

This is the first truth about Wuhan: there is no evidence that SARS-CoV-2 naturally emerged.

None.

First, I want to point out some cold hard truths about the Chinese.

In business and international politics, the Chinese lie, cheat, and steal every step of the way through a business deal or negotiation. In China, it is socially acceptable to engage in these behaviors and is expected in their culture and society.

For westerners to think for even a microsecond that the Chinese government, which is controlled by the communist party, would just hand over any information or data, of any type or form, without reviewing it first, is naïve.

After the data and information related to something as serious and tragic as the emergence of a rapidly spreading and deadly virus were analyzed, the data, information, and analyses would likely be discussed by committees in the highest levels of the communist party to determine: (1) If they should release any information or data; (2) If they release data or information, then how should the information or data be presented; (3) To whom should they present the information and data; (4) What information or data is potentially damaging to the communist party; (5) What information or data is potentially damaging to party goals and objectives; (6) When should the information or data be released; (7) Who should be the person to present the data; (8) What combinations of the answers to these questions will be perceived the most favorably by the community; and, (9) What information or data should be restricted in the context of the previous questions?

These questions represent how communism works in practice, and if you don't understand this, there are many great sources of information about communism. My two favorite pieces of media on the subject are the *Death of Stalin*, which is a hilarious comedy, and the HBO miniseries *Chernobyl*, which is a dramatic miniseries about the man-made nuclear disaster that almost doomed the planet.

There are many similarities between how the Soviet Communist Party responded to the nuclear reactor meltdown and how the Chinese Communist Party responded to the emergence of COVID. I think communism is an inferior political system to capitalism, although I absolutely respect the Chinese people's decision to use communism.

Capitalism has its own weaknesses, and the United States is on the precipice of becoming a full-fledged crony capitalist state, which is only good for the ultra-wealthy or government leaders and can lead to the same kind of decision and policy making that happens in socialist or communist states, with the objective being the consolidation of capital and power.

For those of you not familiar with the term, *crony capitalism* is an economic system in which businesses thrive not because of free enterprise, but rather as a return on money amassed through collusion between a business class and the political class.

Does this remind you of anything that you just read?

I hope it does, because crony capitalism played a significant role in creating the perfect storm of an environment for EcoHealth to rise to power as a middleman.

Drs. Peter Daszak, Billy Karesh, Jon Epstein, Kevin Olival, and I were always socializing and strategizing with global key players at powerful companies and in governments. Our objective was to create an alliance of companies, organizations, and government officials that dominated the sources of funding and controlled the entire scientific discourse within the field of zoonotic emerging infectious diseases.

With this domination, EHA could submit proposals to government agencies that had errors, weak or unfounded methods, flawed assumptions, or overpromises the impact the project would have to fuel the EHA hype machine. This becomes highly problematic when the people that need to be critical of and objective about your work are subject to the power you or your organization wield.

This, in my opinion, was how EcoHealth Alliance was able to bypass the ban on dual use research of concern (DURC), infectious disease research that can be used for peaceful or harmful purposes, in the United States, and that was likely part of why "Understanding the Risk of Bat Coronavirus Emergence" was funded when it was clearly GoF and DURC, and it could be reasonably argued that EHA set up China to fail. If you are a more powerful scientific organization and possess more advanced capabilities than one of your consortium partners or subcontractors, then it is in your organization's best interest to ensure that they are set up for success.

In this case, that would mean leading by example.

EHA should have had the proper internal biosafety risk management framework and a biosafety or environmental health and safety officer with enough of their time dedicated to operationalizing biosafety, biosecurity, and biorisk management, and biosafety and biosecurity programs should have been created, formalized, and refined over time.

Training in these key areas should have been more structured, hands on, and occurred on a routine basis both internally for employees and externally for sub-awards with routine and systematic evaluations and inspections, especially of high-risk steps in the GoF DURC process. This would reduce the likelihood of subcontractors having a catastrophic accident and is simply a good business practice.

With the biorisk system in place, the biosafety officer would then work with the organization's partners to evaluate their risk management framework, insist on inspecting and evaluating their programs in action, and establish a collaborative training program to mitigate any identified deficiencies in a positive and supportive manner.

Maybe the Chinese were opposed to inspections or evaluations by EHA? I strongly suspect that, in 2014, EHA likely could not afford hiring a biosafety officer (BSO) and the expenses associated with their duties. If they could not afford the BSO, then Peter should have made the ethical decision, which is outlined in the Fink Report, to not submit the proposal for the GoF and DURC work to NIH.

While offering these suggestions now is a little too late for humanity, the intent is to show that EHA was in over its head, and this work should not have been approved for funding by NIH NIAID. The proposal submitted by EHA to NIH should have been rejected for a few different reasons.

The first being that EHA did not have an Institutional Biosafety Committee (IBC), which is required for GoF and DURC research.

The second being that EHA did not have a BSO.

And the third being that the select agent report form submitted as part of "Understanding the Risks of Bat Coronaviruses" was grossly inadequate and did not discuss any of the gain of function work described in the Specific Aims. This should have been obvious to peer reviewers if it was peer reviewed.

Fourth, the proposal should have been reviewed by the Recombinant DNA Advisory Committee (RAC) and probably by Dr. Anthony Fauci, if it was not, and it is not clear (at least to me) how EHA's proposal could have gone to the RAC, without having filed registrations with the government for the RAC while not having an IBC or BSO in 2013–2015. An investigation should be launched by the US government to determine exactly how this deficient proposal was approved.

The second truth about Wuhan is that the emergence of the SARS-CoV-2 is clearly not all China's fault.

Peter Daszak unethically submitted a proposal where he intentionally obfuscated the real nature of the work (GoF and DURC). EHA and Peter Daszak were clearly not in compliance in terms of biosecurity and biosafety personnel, programs, and (potentially) registration with the RAC. The NIH or peer review should have rejected the proposal.

Now this is where things get a bit trickier, and there is more speculation throughout this chapter. This upcoming section explores the emerging COVID crisis in the context of how the Chinese government functions.

First, based on China's previous pattern of behavior, it is safe to say that it is in China's best interest to control the narrative, as the preponderance of scientific evidence strongly suggests the origin did, in fact, occur in China.

Since China wants to control the SARS-CoV-2 narrative, we can hypothesize what their goals of the narrative could be:

1. to demonstrate power to the world
2. to reduce fear in their own country
3. to avoid military conflict due to releasing a bioweapon intentionally, when they did not release a bioweapon intentionally
4. to avoid military conflict when they released a bioweapon intentionally
5. to be perceived as in control
6. to appear superior to their competitors
7. to help the world
8. to create chaos and confusion internationally
9. to undermine opposing narratives

10. to consolidate and increase power within the communist party
11. to exert China's power globally
12. to gain competitive advantages
13. to encourage regional or global peace and stability

The strategic objectives outlined are not mutually exclusive, and some can be eliminated immediately based on analysis of China's behaviors in late 2019.

- Number 1 can be confirmed, as the behavior and messaging exuded realistic power and confidence.
- Number 2 can be eliminated as citizens of China were initially reporting their fear, and videos that were posted widely showed panic and fear on the streets of China, which the Chinese government attempted to suppress.
- Number 3 can be eliminated immediately. If this were a strategic goal, then the best course of action is to immediately share all your data and to accept foreign medical or investigative help, and the opposite behavior occurred.
- Number 4: by delaying investigation and not releasing data rapidly, it would make an investigation into the origin of COVID difficult; however, we now know what China very likely knew then, that SARS-CoV-2 was made in a lab. Therefore, this does not eliminate an accidental laboratory release or an intentional release, because the behavior indicated that China wanted this to look like a naturally emerging disease when they decided to communicate the problem to the world.
- Number 5 is confirmed as China acted quickly and decisively, but not necessarily in alignment with when the disease first emerged.
- Number 6 is supported. Messaging in China throughout 2020 emphasized how generous the Chinese Communist Party (CCP) and Chinese government are and that their generosity can bring calm to a chaotic world.
- Number 7 is supported. By waiting months to report the outbreak in late December 2019, the investigation into the origin of COVID was intentionally obfuscated.
- Number 8 is supported. By waiting months to report the outbreak in late December 2019, the speed with which the outbreak expanded

globally made the transmission rate appear worse than it was to public health experts.

- Number 9 is supported; China immediately began campaigning that the United States intentionally released the agent in China.[231] [232] [233] [234]
- Number 10 is supported: The CCP messaging demonstrated that it wants to reassure everyone that it is the only party necessary.
- Number 11 is supported. The CCP used COVID to increase technological surveillance capabilities and party membership increased at a rate of two million new members per year.
- Number 12 is supported;[235] [236] [237] [238] China's massive investments in surveillance technology and machine learning benefit all sectors of China's economy.
- Number 13 is inconclusive; China is engaging in global peacekeeping missions and assuming a larger role in creating international stability in resource rich, but financially poor, countries. Many view this as China expanding its military globally and that it has been combined with rapid modernization.

From examining the behavior of the CCP and China from 2021–2022, it potentially can help us infer what their goals and objectives were in relationship to the emergence of COVID. This can help develop scenarios which can be qualitatively or quantitatively analyzed to determine what likely happened in China.

I think it is safe to say that China would be terribly embarrassed by a laboratory leak which resulted in the deaths of over six million people.

Since China knew from day one that this was a genetically engineered agent, which I also immediately knew, the narrative and behavior encouraging the "wet market" emergence theory in Wuhan did not make sense.

The Chinese government and the CCP have used advanced technology, censorship, and data manipulation across a wide variety of issues. There is no reason to suspect that the Chinese would behave any differently here. Simply, if one wanted to manipulate the entire scientific narrative around an infectious disease outbreak, you would plot the locations of each case in a geographic information system, or in a database, by the diagnosis date, and then remove the case data from the table or database that did not fit the narrative that one wanted others to find and report.

This probably wouldn't be any more than a half-hour to an hour exercise for most people in epidemiology, data science, or biostatistics.

This is exactly what the Chinese did, and perfectly fits their *modus operandi.*[239] [240] [241] [242] [243]

Then, they handed the dataset off to the world and told everyone that the dataset contained all the records, and barely any of these "famous" or powerful scientists challenged the data. Instead, they did what academics typically do: they grabbed the data as fast as they could, analyzed the data in various ways, and submitted works to publications as fast they could. Voila! The "wet market" was the location from where SARS-CoV-2 emerged.

About that "Wet Market"

A "Live Animal Market" is a type of market where live animals are slaughtered and processed when purchased. Often, wet markets are in resource poor countries, and have poor sanitation and environmental health. The animals are typically hot, stressed out, and have poor nutrition, which weakens the live animals' immune systems. In live animal markets, live animals are stacked on top of each other in cages where they defecate, urinate, or swap saliva or other biological materials with each other.

"Wet Markets" sell consumption-oriented, perishable goods in a non-supermarket setting. These markets were named after their frequently wet floors, a result of regular cleaning to keep food stalls sanitary and the melting of ice to keep foods fresh.

Sometimes wet markets have a small selection of live animals for sale, and they only represent a small fraction of the food products available. Immediately, when they identified the "Wet Market" as the source of the outbreak I went to my computer and started to look at pictures of the establishment and the surrounding neighborhood, and it appeared to be a clean and sanitary market, packed full of seafood, and the only pictures of live animals that I could find were stored hygienically.

The neighborhood looked modern, clean, and upscale, akin to the upper east side of New York. Then, I found the name of the establishment: the Huanan Seafood Wholesale Market. The idea that SARS-CoV-2 came out of such a nice and clean market was ridiculous.

The market did not have any of the conditions that I described in the definition of a live animal market, which are the types of conditions typically necessary for novel emerging infectious diseases typically arise.

The mainstream media was implying that the Huanan Seafood Wholesale Market was a live animal market, even though they are different from wet markets, which are typically not a great source of emerging infectious disease risk, especially in the pristine condition which the market in question is kept.

What is amazing to me is how many scientists have locked onto the narrative that the Huanan Seafood Wholesale Market was the source of the outbreak, yet hundreds of other global case reports and analyses of human behaviors, which are highly correlated with disease outbreak patterns, completely undermine this narrative which was created by China and spoon fed to US officials, their research partners, and the mainstream media. They were all too eager to support the narrative, some using pseudoscientific techniques, which were published in countless peer-reviewed journals.

Then, the US government and mainstream media pointed to these peer-reviewed pseudoscientific works of art that perfectly aligned with the official Chinese and US narrative. Yet, nobody is challenging these scientists on the foundational assumption of the validity and reliability of the data.

An honest scientist would say that they couldn't possibly know their validity and reliability.

The global case reports from clinical diagnoses or diagnostic tests indicate that SARS-CoV-2 emerged in August 2019.

Other case reports support emergence dates of October and November 2019.

These dates are well before the late December 2019 timeline that was communicated to the public in the United States, and I have good reason to believe that the US government was alerted to the outbreak in August or October 2019.

When Dr. Amy Jenkins from DARPA contacted me in October 2019 to recruit me to be a program manager in biologics and emerging infectious disease work, I was flattered. I also wondered how she had found my brand new cell phone number, and I doubted that she remembered the first time we met at the University of Minnesota when I was a PhD student.

Being recruited to be a program manager at the world's premier advanced technology development institute had been my dream job just a few years earlier. When she contacted me, I was making incredible money, and I did not want to get a top-secret security clearance again, nor did I want to move to the Washington, DC, area while also taking a significant reduction in salary.

I politely declined the offer, and she asked me to sleep on the decision and speak again the next day, to which I agreed. That night I went home and told Emily about the opportunity, and she felt the same way I did about all aspects of the proposition.

So, when Amy called the next day, I once again politely declined the offer and thanked her.

I never thought much of it until a few months later when COVID emerged.

About a month after the pandemic began, and I was adamant that SARS-CoV-2 was a manmade agent, I suddenly realized what Dr. Jenkins's potential motivation and persistence for recruiting me were.

The intelligence community (IC) realized that I was the only person in a senior position that had left EHA, and the fact that I was working outside the government's control made me a threat to their agenda.

I surmised that the IC, DoD, DIA, or CIA thought there was a high probability that I would accept the offer for my former dream job.

Then, after I received the top-secret security clearance, they would "read me in" to the biological program, and then I would have been sworn into silence for the rest of my life.

If my hypothesis is correct, wow, they are clever!

The only problem was that Dr. Jenkins contacting me randomly later raised my suspicions in 2020 that the US government or DoD likely knew about the SARS-CoV-2 outbreak before October 2019.

I was a loose end at EHA. Everyone else working in management at EHA is still employed at the company. Maybe this was just a coincidence? I find it highly suspect as I can't fathom how or why I was contacted for this position as I had told everyone in my professional national security sphere that I no longer wanted to work in the classified space.

So if the government knew about COVID-19's emergence in the early fall of 2019, then why didn't they warn the public and tell everyone the truth?

There are many different types of biological agents that can be used as weapons. Each biological agent has an optimal delivery method to maximize lethality and dispersion. Typically, the ideal bioweapon is one that has high lethality, disperses broadly, and transmits easily.

The problem with bioweapons is that their use is easily noticeable, and they are difficult to deploy. For these reasons, biological warfare agents like smallpox, anthrax, or hantavirus have less appeal in modern day asymmetrical warfare. Bioweapons have more appeal for terrorists as they are cheap and a small, targeted attack can create widespread panic and fear.

While bioterrorist attacks can cause widespread panic and harm, the duration of fear is often short-lived unless there are repeated attacks, like the series of anthrax attacks in 2001.

The fear generated by a single attack quickly fades, unlike the long duration of fear associated with the COVID pandemic.

The COVID pandemic generated unwarranted hysteria across the globe for two years. Even I was walking around with a full-face respirator for the second half of January in 2020. Initially, the fear that COVID caused was mostly associated with the uncertainty associated with a novel virus.

Later in the pandemic, the fear associated with COVID was mainly due to misinformation and fear mongering.

COVID causes a wide range of symptoms in the afflicted, with a wide range of disease severity. While the disease is mild to moderate in severity for most cases, people with risk factors like asthma or obesity are at much higher risk of having severe disease.

COVID is also highly transmissible, but the transmissibility of COVID could decline over the next two years as more people in the population acquire immunity to the disease. Only time will tell what the endemic transmission rate of COVID will be. This begs the question: Why would a nation use COVID as a bioweapon?

Traditionally biological weapons are used to demoralize, incapacitate, force the use of vast medical resources in treating the wounded, or to tactically deny an enemy's ability to occupy or move through an area.

SARS-CoV-2's use as a bioweapon is appealing because it is easy to disseminate, causes tremendous fear, can overwhelm health-care systems, and it is stealthy. As a potential bioweapon, the disease at first presents like many other upper respiratory infections and would likely go undetected for a long enough period. Thus, the disease could continually transmit beyond a point where it could be detected and contained.

If you wanted to disrupt large populations for a prolonged period while causing economic disruptions, then SARS-CoV-2 might be highly effective in achieving these objectives. I say "might" because global exposure to SARS-CoV-2 has resulted in a substantial portion of the population being immune to the disease. Consequently, it might only have been effective as a single use bioweapon.

To deploy SARS-CoV-2 a second time as a bioweapon, it would require genetic modifications to evade the global population's acquired immunity, as research indicates that naturally acquired immunity to SARS-CoV-2 is robust.[244] [245] [246] Only a very limited set of nations have the training and laboratory capacity to successfully execute this type of advanced research and development.

To get to the truth about whether SARS-CoV-2 was leaked or intentionally released, I like to use scenario analysis to determine which scenarios are the most likely. Then, I use qualitative or quantitative methods to make objective comparisons between the scenarios.

As before, the first step is developing or imagining all the potential scenarios and trying to eliminate as many as possible with logic.

- Scenario 1: The Wuhan Institute of Virology (WIV) accidentally leaked SARS-CoV-2 one time.
- Scenario 2: The WIV accidentally leaked SARS-CoV-2 multiple times.
- Scenario 3: The Chinese intentionally released SARS-CoV-2 one time.
- Scenario 4: The Chinese intentionally released SARS-CoV-2 multiple times.
- Scenario 5: The United States or other state-sponsored actor released SARS-CoV-2 in China.
- Scenario 6: First, China intentionally released SARS-CoV-2. Second, the WIV accidentally released SARS-CoV-2.
- Scenario 7: First, the WIV accidentally released SARS-CoV-2. Second, China intentionally released SARS-CoV-2 as a cover operation to conceal the accidental release.
- Scenario 8: Rogue and criminal actors within the US government worked with the Chinese government to intentionally release the agent multiple times.

Perhaps you can think of different scenarios than I did? If so, then you might have solved the puzzle. Immediately, a few scenarios.

- Scenario 1 can be eliminated because there is substantial evidence that the virus was spreading globally as early as September 2019, and that the different genetic lineages of SARS-CoV-2 circulating wouldn't likely have been due to genetic drift and shift.
- Scenario 2 is plausible and is supported by the scientific data.
- Scenario 3 can be eliminated because there is substantial evidence that the virus was spreading globally as early as September 2019, and that the different genetic lineages of SARS-CoV-2 circulating wouldn't likely have been due to genetic drift and shift.
- Scenario 4 is plausible and is supported by the scientific data.

- Scenario 5 can be eliminated because it would have likely caused a war, and there would have been no reason for China to delay notifying the world what really happened. An intentional release by a foreign actor in China would have been immediately used as propaganda by the CCP.
- Scenario 6 is not plausible and can be eliminated.
- Scenario 7 is plausible as these actions by China would provide excellent cover for an accident, and intentionally releasing an agent at the seafood market in Wuhan would create the data necessary to point to the market.
- Scenario 8 is not plausible and can be eliminated as it is highly unlikely that the US administrative officials involved would have been able to leave no trace. The US cover-up of its involvement in the development of SARS-CoV-2, led by Dr. Anthony Fauci, would have not been so haphazard, which left a massive incriminating digital and paper trail.

Unfortunately, this logic-based exercise has not eliminated enough of the scenarios. The next way that we can attempt to eliminate scenarios is to think about the potential motives for an intentional release of SARS-CoV-2 in the social and political context that existed when SARS-CoV-2 emerged in September 2019.

Intentional Release Motive 1: Global Corporate Greed and Money

One of the leading motives that has been discussed among conservative, libertarian, and free-minded critical thinkers alike is the Great Reset. The Great Reset is not some conspiracy theory, and the intent, motive, and goals of the program have been developed and promoted by the World Economic Forum (WEF) and its chairman Klaus Schwab (that in my opinion, looks like a penis).

Even I was skeptical at first that this could be a plausible motive for an intentional release of SARS-CoV-2 until I watched Klaus Schwab speak about the Great Reset and read the piles of documents published by the WEF.

Perhaps the release of SARS-CoV-2 was merely a lucky coincidence for the WEF's stated goals and objectives? Klaus Schwab has been pushing the concept of "stakeholder capitalism" for decades.

The idea is that global capitalism should be transformed so that corporations no longer focus solely on serving shareholders but become custodians of society by creating value for customers, suppliers, employees, communities, and their other "stakeholders."

The WEF envisions stakeholder capitalism being carried out through a range of "multi-stakeholder partnerships" bringing together the private sector, governments, and "civil society," whatever that means, into a single global governance. This means giving corporations more power over society and democratic institutions less power. The WEF's vision is "the government voice would be one among many, without always being the final arbiter."[247]

For Americans, that means that the US government would be just one stakeholder in a multi-stakeholder model of global governance.

Who are these other, non-governmental stakeholders?

Large corporations.

The WEF, best known for its annual meeting of high-net-worth individuals in Davos, Switzerland, describes itself as an international organization for public-private cooperation. WEF partners include some of the biggest companies in oil (Saudi Aramco, Shell, Chevron, BP), food (Unilever, the Coca-Cola Company, Nestlé), technology (Facebook, Google, Amazon, Microsoft, Apple), and pharmaceuticals (AstraZeneca, Pfizer, Moderna). Instead of corporations serving many stakeholders, in the multi-stakeholder model of global governance, corporations are the official stakeholders in global decision-making, while governments are relegated to being just another minor stakeholder.

Corporations become the main stakeholders, and governments must negotiate with them.

If this were not terrifying enough, they have something called the COVAX initiative.

The COVAX initiative aims to "accelerate the development and manufacture of COVID-19 vaccines, and to guarantee fair and equitable access for every country in the world."[248] COVAX was set up as a multi-stakeholder group by two other multi-stakeholder groups: Gavi, the Vaccine Alliance, and the Coalition for Epidemic Preparedness Innovations (CEPI).

Both Gavi and CEPI have strong ties with the WEF, and the WEF was one of the founders of CEPI. Gavi and CEPI have strong ties to the Bill and Melinda Gates Foundation, Pfizer, GlaxoSmithKline, AstraZeneca, Johnson & Johnson, and the large chemical companies that manufacture the vaccine's ingredients, through manufacturer partnerships (Gavi) or as "supporters" (CEPI).

Even though COVAX is funded predominantly by governments, it is these corporate-centered coalitions that are overseeing its roll-out. The contrast between the multi-stakeholder approach and a classic multilateral one came to the surface when South Africa and India proposed the so-called TRIPS waiver at the end of 2021.

They requested a temporary lifting of intellectual property rules on all COVID-19 technologies to boost the manufacturing and distribution of vaccines and other essential medical products in mainly developing countries.[249] [250]

This would have significantly reduced the profitability of the companies listed previously and the many other companies that I omitted.

WHO Director General Tedros Ghebreyesus said in a speech that he backed the proposal.[251] However, Gavi, the Bill and Melinda Gates Foundation, Bill Gates, and Big Pharma opposed the TRIPS waiver.[252] [253] Clearly, the main goal of these entities is to maintain their profitability, rather than protecting people from COVID.

As crazy as it might sound, the WEF, Big Pharma, and the chemical industry had the motive to release SARS-CoV-2 to fulfill their policy agenda.

Intentional Release Motive 2: To Damage President Donald Trump

The other leading motive for an intentional release is that China wanted to damage President Donald Trump's chances of being re-elected.

In September 2019, President Trump's economy was the strongest in US history, and his economic trade policies aimed at leveling the playing field with China were highly effective.

For the first time in decades, the United States was gaining leverage over China in trade, and President Trump was promising to double down on his tough trade policies toward China if re-elected.

By intentionally releasing SARS-CoV-2, Donald Trump's political opposition would be able to question his effectiveness as a leader. The hope for China would be that by undermining President Trump that a weaker leader that was more sympathetic to China would be elected.

Now we can revisit the intentional release scenarios and determine if any can be eliminated with context of the obvious motives:

- Scenario 4 is supported by the motives of the intentional release scenario.
- Scenario 7 is not supported by the motives and cannot be eliminated, as the primary cause of this scenario begins with an accident.

Now let's revisit the scenarios and see which ones are remaining. The remaining scenarios are:

- Scenario 2: The Wuhan Institute of Virology accidentally leaked SARS-CoV-2 multiple times.
- Scenario 4: The Chinese intentionally released SARS-CoV-2 multiple times.
- Scenario 7: First, the WIV accidentally released SARS-CoV2. Second, China intentionally released SARS-CoV2 as a cover operation to conceal the accidental release.

We are down to three scenarios out of the eight initial scenarios. Next, let's examine some key facts related to the emergence of COVID-19. There are numerous reports from China that they were scrambling to respond to the rapidly spreading outbreak in China in the early fall of 2019.[254] [255] [256] [257]

Business leaders in China were reporting to their government that they were under severe stress due to the pandemic.

The WIV was scrambling to procure BSL4 containment equipment.

There were personal protective equipment (PPE) shortages in China.

Also, athletes that attended the 2019 Military World Games, which were held from October 18–27 in Wuhan, reported that the city was a ghost town. Let's revisit the scenarios.

- Scenario 4 can be eliminated. SARS-CoV-2 was spreading globally prior to the Military World Games, and this would have been the best place to release the agent. Moreover, it would have been easy to target foreign athletes and expose them to the agent as they were returning home, and by this time the epidemic in China was well under way. Additionally, if there was an intentional release, there would be no justifiable reason that I can envision where releasing the agent in your own country would be tactically beneficial. The Chinese certainly have the resources to transport an agent to a foreign country where a person or a group of people could infect themselves, and then spread the agent. The would be the best way to deploy and disseminate SARS-CoV-2 as a bioweapon. In fact, this method of dispersing a highly transmissible bioweapon is the most feared and this would have been the best way to have plausible deniability in an intentional release scenario. China could have just claimed that the disease emerged in a live animal market elsewhere around the globe.

- Scenario 2 cannot be eliminated. For this scenario to be plausible, we would have to believe that the Chinese were initially unaware of the SARS-CoV-2 leak for a long period, perhaps months, without mitigating the threat. Documents from the WIV indicated that they were suddenly buying biocontainment equipment and upgrading the WIV in response to a lab leak which was detected by Chinese officials in October 2019. To not respond and mitigate a lab leak of SARS-CoV-2 when it was detected would be suicidal politically and physically.

This leaves two scenarios remaining, and I believe to be the truth about Wuhan, which is supported by the facts and a very similar pattern of communist behavior that is almost identical to what took place at Chernobyl Nuclear Power Plant on April 26, 1986.

In the years leading up to the disaster at Chernobyl, Soviet officials had covered up several accidents and dangerous situations prior to the 1986 explosion of the reactor. On the date of the disaster, Soviet and Communist party officials spent the first eighteen hours after the reactor explosion attempting to cover-up the fact that the reactor had exploded.

Communist countries focus heavily on propaganda and messaging as individuals in these systems are incentivized to not have any problems, mistakes, or failures. This is because the party often believes that the collective cannot have made mistakes, and thus any mistakes are due to the failure of individuals that were not operating as part of the social collective.

This type of thinking and the associated behaviors lead to catastrophic risks being assumed by managers overseeing dangerous work, and when these catastrophic risks are realized, managers in the communist party going up the chain of command are heavily incentivized to conceal accidents, failures, and disasters.

The evidence surrounding the emergence of SARS-CoV-2 only points to two scenarios:

Scenarios 2 and 7 are supported by all the evidence and both are plausible. SARS-CoV-2 leaked from the laboratory as early as August 2019. The probable cases detected in Italy in 2019 make sense as Italy is one of the most popular tourist destinations of the Chinese. The confirmed cases of SARS-CoV-2 in Italy in October 2019 further support this scenario. By mid-October 2019 the disease was likely already on every continent except Antarctica. The DoD, along with militaries globally, detected the disease in their countries' service members that attended the Military World Games in Wuhan. Trying to avoid global panic, governments began to implement response plans, which included the US government's mobilization of the mRNA SARS-CoV-2 gene therapy. The Chinese and some of their US government collaborators at the Department of State, USAID, and the DoD went into full cover-up mode. This is evidenced by CIA operatives working the COVID cases in China, like Dr. Michael Callahan, to collect valuable intelligence on the diseases. The Chinese worked ferociously to contain the spread, and likely slowed down the spread to some degree throughout November 2019.

In December 2019, the Chinese government potentially intentionally released the agent at the Huanan Seafood Wholesale Market, which would be necessary for the epidemiologic investigation of the disease spread to focus on the market. Without having an intentional release at the market, the spatial and temporal pattern of cases would not point to the market as a point source. This would account for the differences in genetics observed in SARS-CoV-2 measure in early cases. This last scenario assumes that Chinese didn't fabricate, censor, or curate the entire dataset provided for analysis, which is certainly possible, and is just as nefarious as intentionally releasing the agent. Clearly, the agent was released in the market or all the epidemiologic case data was censored or manipulated related to it, which is quite plausible.

This leads to a shocking realization about the United States' response to the COVID pandemic. Nobody should be surprised that the Chinese lied about the outbreak of SAR-CoV-2 and then went to extraordinary lengths to make it appear as if the disease naturally emerged.

The shocking part of all of this is how the United States government lied to all of us.

Several agencies were aware that this outbreak was occurring and did not communicate this disaster to the public.

I find it more concerning that this emerging threat was not communicated to President Donald Trump.

This should not have come as a surprise to him in early January 2020.

It begs the question, was this depriving the president of vital national security information intentional, or was President Trump informed months earlier and chose not to act? Only he and his closest advisors know the answer to those questions.

People have pointed to President Barack Obama's executive order lifting the ban on gain of function research a week before he left office in 2016 as evidence that SARS-CoV-2 was a dirty political trick by President Obama.

The reality was that the gain of function work on select agents had never actually stopped.

Dr. Peter Daszak circumvented the GoF DURC policy with who appears to be his accomplice, Dr. Anthony Fauci.

Only an in-depth investigation as to how the proposal for the project titled "Understanding the Risk of Bat Coronavirus Emergence" was approved, since it contained DURC and GoF, will tell us who in the US government is to blame for the transfer of dangerous biotechnology to the Chinese, in addition to Drs. Fauci, Daszak, and Ralph Baric.

CHAPTER 18

Biggest US Intelligence Failure since September 11

The COVID story represents the largest intelligence failure in US history.

The intelligence community (IC) receives intelligence via a variety of sources and methods. The disciplines of the intelligence community are human intelligence, signals intelligence, imagery intelligence, measurement and signatures intelligence, and open-source intelligence.

The CIA likely had at least two people collecting human intelligence on China, Dr. Peter Daszak and Dr. Michael Callahan, based on Peter's statements to me being accurate.

I had developed some of the signals intelligence capabilities for the IC, including the DoD, which would have easily detected the early digital signals of an emerging infectious disease outbreak.

Personally, I had used a combination of open-source, measurement and signatures, and imagery intelligence to analyze air pollution in Wuhan, which indicated that bodies were being cremated at abnormally high levels in mid-December 2019, and I confirmed in late 2020 that other intelligence analysts had independently used similar techniques and drew the same conclusions about an earlier SARS-CoV-2 outbreak than what was being communicated by Chinese and US officials.

We know that, as early as January 2018, State Department officials issued warnings about the WIV, including the serious shortage of appropriately trained technicians needed to safely manage research on potentially deadly zoonotic coronaviruses.

In my experience, when there is one intelligence report that the government released publicly, there are likely at least ten more that were similar and unreleased, as they would want there to be high confidence in the report's accuracy, before releasing it publicly.

If I was worried about laboratory biosafety and biosecurity in foreign laboratories funded by EcoHealth Alliance, then it appears that I was not the only one, and I have never stepped foot in China or the Wuhan Institute of Virology.

Lastly, the DoD must have detected that some of their service members that attended the World Military Games in Wuhan had contracted COVID, and as a matter of operational security this illness would not be reported publicly.

However, if service members did become ill at the World Military Games, then public health authorities in the United States should have been notified or warned, as an emerging public health threat like COVID is a threat and concern to all Americans.

With all the intelligence that the US government had to the biothreats and bio risks associated with the Wuhan Institute of Virology, and the fact that at least the CIA/DIA, DoD, and NIH knew about the GoF DURC work funded by NIAID, why wasn't this information communicated to the American people, and perhaps the US president, sooner?

Therefore, this is the biggest intelligence failure in US history.

CHAPTER 19

Biggest Cover-Up in History

The cover-up of SARS-CoV-2 began with the Chinese in September 2019, and this fact should not be surprising.

If communist superpowers exist, then the people that live and work in those systems will be incentivized to protect the party at all costs. The emergence of SARS-CoV-2 is the second time in history where a massive cover-up operation was deployed to conceal the true nature and extent of a disaster.

The US cover-up of SARS-CoV-2, which was also likely supported by some of its closest allies, likely began in October 2019, as this was the same time DARPA attempted to blindly recruit me for a position that I was not vaguely interested in at the time.

I am guessing that the intelligence community and the DoD were hoping that the disease would burn out, which often happens with emerging infectious diseases, and they were probably not aware of the extent to which SARS-CoV-2 had already spread globally.

Even so, I would suspect that they would have provided classified briefings to leadership at HHS, CDC, NIH, and FDA, but that appears not to have happened based on the public statements from CDC Director Robert R. Redfield (and others), where his public statements did not confidently name a laboratory leak in China as the source of the outbreak.

I find all these facts quite perplexing.

Less puzzling to me is how the IC seems to be split to this day as to whether the source of the SARS-CoV-2 outbreak was either naturally emerging or the result of a laboratory leak.

The IC's mission is to protect the United States, and by casting doubt over the origin of the outbreak, and by the United States aiding the Chinese in the design and engineering of SARS-CoV-2, a split analytical report published by the Directorate of National Intelligence, where half the analysts conclude that it is a lab leak and the other half conclude that SARS-CoV-2 naturally emerged achieved this goal by further muddying the water so that a final attribution of cause and effect cannot be determined.

This type of behavior will have the oppositive intended effect of protecting the United States, because by not acknowledging our country's mistakes related to dangerous GoF DURC research and policy, these mistakes are likely to happen again.

In January 2020, Dr. Kristian Andersen of the Scripps Research Institute had been examining the genetic characteristics of SARS-CoV-2. While I worked at EcoHealth, Dr. Andersen and I had been looking for ways to collaborate. In an email exchange with Anthony Fauci and Jeremy Farrar (Wellcome Trust), Andersen stated:[258] [259]

> The problem is that our phylogenetic analyses aren't able to answer whether the sequences are unusual at individual residues, except if they are completely off. The unusual features of the virus make up a really small part of the genome (<0.1%) so one has to look really closely at all the sequences to see that some of the features (potentially) look engineered. … all find the genome inconsistent with expectations from evolutionary theory …. there are still further analyses to be done, so those opinions could still change.

Just four days later, Andersen gave feedback in advance of a National Academies of Sciences, Engineering, and Medicine letter that was referenced in the prestigious *The Lancet* medical journal to argue *against* the idea that the virus had been engineered and brand it a conspiracy theory.[260] [261]

Sound familiar?

Dr. Andersen called the idea that the virus was engineered "crackpot theories," stating that "engineering can mean many things and could be done for basic research or nefarious reasons, but the data conclusively show that neither was done."

So, I decided to examine all of Dr. Anderson's funding from NIH, and you will never guess what I found.

Dr. Andersen's funding from NIH and NIAID dramatically increased after he reversed his position that SARS-CoV-2 had all the signatures of being engineered.

In fact, his funding in 2020 increased at a rate that I have never heard of or seen in the field of emerging infectious diseases research. His "continuing funding," a statistic used by government agencies that fund research, nearly triples from $7,141,011 to $23,724,681.[262]

This is the corruption that exists within our research entities and is emblematic of the lengths that the US government has gone to cover up the truth about its involvement in the development of SARS-CoV-2. Do you recall what I said about the Wellcome Trust's relationship with EcoHealth Alliance? The Wellcome Trust was in it for access to the valuable genetic samples that are necessary to build medical countermeasures, therapeutics, or highly profitable new mRNA injections that are now defined "vaccines."

Why was the president of the Wellcome Trust, Jeremy Farrar, included in this email exchange between Fauci and Andersen?

I have some ideas as to why, but only a formal investigation will provide us with the real reason.

CHAPTER 20

I Am Over the Target

Never in a million years did I think that my own government would launch an illegal psychological operation against me.

Their goal was to discredit me, to scare me, and to prevent me from speaking out.

Then, their goal became to prevent the publication of my book, and they failed.

I collected piles of cold, hard evidence of the crimes committed against me by the US government, including photographic evidence of trespassers and highly sophisticated drones being flown at distances of a few feet from my house caught on camouflaged motion activated trail cameras, and a witness list that is two pages long.

The witness list includes my attorney, Thomas Renz, who heard a drone hovering over my house during a visit to record a deposition under oath with penalty of perjury, where I covered much of what is written in this book.

Next is a list of what has happened to me since late October 2021. The US government is terrified of the truth.

This list was taken from the Temporary Restraining Order and Injunction that Renz is preparing to file against the US government and the Michigan State Police:

Dr. Huff has alleged facts in a whistleblower action before (Senator Gary Peters, Senator Ron Johnson, DHS, HHS, DoD, DTRA, USAID, OIGIC, DARPA, USDA) and in multiple published news articles, which demonstrate that the US government has wasted billions of dollars of taxpayer funding and is partially responsible for the development of the biological agent known as SARS-CoV-2, the agent that causes the disease COVID-19. The SARS-CoV-2 agent was partially developed in the United States with US taxpayer funding, and Chinese researchers from the Wuhan Institute of Virology received the necessary and critical training related to viral gain of function research in the United States. The forthcoming facts also suggest that the emergence of SARS-CoV-2 was also a US government intelligence failure. Lastly, these facts will also demonstrate that the agent known as SARS-CoV-2, and the SARS-CoV-2 mRNA vaccine, originated at a laboratory at the University of North Carolina, before the SARS-CoV-2 outbreak in China occurred. A timeline of the related scientific milestones, related parties, and Dr. Huff's personal knowledge of the facts in support of his whistleblower allegations is attached.

Dr. Huff is experiencing severe harassment in retaliation for his whistleblower action from government actors as described herein.

In October 2021, a man by the name of Mike Bennett contacts MTRX INC and speaks with MTRX's Chief Revenue Officer Andrew Duso. Bennett pretends to be interested in purchasing or learning more about MTRX INCs products and asks Duso to speak with Huff. Huff agrees to speak with Bennett about MTRX INC technology. After pitching the technology to Bennett, Bennett pivots the conversation to the origin of COVID and Huff's time at EcoHealth Alliance. Bennett claims to be a US Navy veteran and a doctor. Huff assesses that these could be accurate claims and that Bennett has the most inside knowledge of EcoHealth Alliance, the USAID PREDICT program, and the gain of function work funded by the US government, of anyone that he has ever met that did not have first-hand work experience in these programs. Huff immediately concludes that Bennett is a CIA operative based on Huff's past interactions with the CIA and the Intelligence Community and communicates his belief directly to Bennett. Bennett states that he has no connection to the federal authorities. Huff does not believe Bennett and continues to communicate with him, despite Huff's belief that he is a CIA agent, to ascertain what Bennett's motives are.

On October 26, 2021, Huff is contacted by a man named Adam Shore on LinkedIn. Huff speaks with Shore and believes that Shore's motives could be to distract Huff.

On October 26, 2021, Huff is contacted by a man named David Lopez on LinkedIn. Lopez claims to be a former Navy Seal and claims that he is involved in numerous strange businesses including building a community on an island to support masculinity and manhood and is heavily tied to alternative cryptocurrencies and technologies like blockchain. Huff decides to continue speaking with him. Lopez introduces Huff to the producer of the film *Plandemic*, Dr. David Martin, and offers to help get him on Fox News' *Tucker Carlson Tonight* to tell Huff's story. Huff later concludes that he is a federal agent or security contractor, which is independently verified by an online acquaintance running for US congress in the State of Texas. This person refers to him as a "white hat." As of October 26, 2021, Huff has not come forward publicly as a whistleblower.

On October 27, 2021, Huff receives a fraud notice on his USAA credit card. Huff immediately contacts the bank. The credit account could not be found by the bank's staff. The account was deleted from the bank's records.

On October 27, 2021, Huff exchanges text messages with Giles Demaneuf. Giles is a prominent data scientist and member of an open online community named DRASTIC, which has been investigating the origins of COVID. Huff believes Demaneuf to be a genuine and good actor.

On the night of October 27, 2021, Huff speaks with his wife Emily on the FaceTime app and tells her to immediately change her passwords on all her accounts. Huff has not yet come out publicly as a whistleblower.

On October 28, 2021, Huff begins to receive numerous text messages warning him to stay safe, as well as indirect death threats, via social media platforms like Twitter and LinkedIn. Huff has not yet come out publicly as a whistleblower.

On October 28, 2021, Huff exchanges text messages with Patrick Davis. Davis is a member of the staff from the House Permanent Select Committee on Intelligence for the Minority, Republican Party. Huff and Davis begin to discuss what Huff knows about EcoHealth Alliance and the origin of COVID. Huff has not yet come out publicly as a whistleblower.

On October 28, 2021, Huff exchanges text messages and converses with Jan Jekeliek, editor-in-chief at the *Epoch Times*, related to doing a sit-down video interview related to Huff's story. Huff has not yet come out publicly as a whistleblower.

On the evening of October 29, 2021, Huff contacts the Sheriff's department to request an emergency concealed pistol license due to the continual escalation of security concerns. The staff is supportive and is, understandably according to Huff, hesitant to bother the Sherriff after normal hours. Huff has not yet come out publicly as a whistleblower.

On the night of October 29, 2021, high end commercial or military grade drones appear at their property at night while the Huffs are sitting in their hot tub, which has a 200-degree view to the south partially obstructed by trees. The conditions at the time were low ambient starlight with good ground to air visibility. Andrew Huff has Emily Huff retrieve a high-powered scope and binoculars to examine the drone and has Emily immediately contact 911. The Huffs are terrified based on the context of the past week's events. Since Andrew Huff has worked with drones as a scientist and became familiar with their capabilities as a student in security technologies, as a senior member of the technical staff at Sandia National Laboratories, and while serving in the US Army, Andrew Huff instructs Emily Huff to duck behind a very thick, dense pile of wood to determine if the drone is tracking their position. The drone maneuvers to maintain line of sight on their position as the Huffs moved between large and dense piles of wood, which suggests that they were in fact the target under observation, and that the drone had sophisticated optical capabilities like night vision or thermal vision. Huff estimates the size of the drone to be five to eight feet in diameter with a quadcopter-like design with multiple rotary wings. The drone is flying higher in altitude than that of consumer grade drones and is above three thousand feet at one mile south/southeast from the house.

Beginning on October 29, 2021, drones are observed flying over the Huff residence each day during daylight hours and from dusk to dawn except for inclement weather.

On the night of October 29, 2021, after reporting the drone to law enforcement, and fearing grave security concerns, Huff decides to immediately come out publicly as a material witness on both Twitter and LinkedIn and publicly states that Dr. Peter Daszak from EcoHealth Alliance stated that he was working with the CIA, and other relevant facts related to fraud and gain of function work conducted by EcoHealth Alliance, which are later specified and submitted in Huff's whistleblower complaints.

On the night of October 29, 2021, after coming forward, Attorney Ravi Batra, a lawyer from New York City who was suing EcoHealth Alliance for creating a bioweapon, offered to represent Huff pro bono. Batra has strong ties to the Democratic Party, Democrats on the US House Intelligence Committee, and personally names Congressman Jerry Nadler as a close friend and regular acquaintance while speaking with Huff on the telephone. Batra claims that his representation will help take the pressure off Huff. Over the next few months, Huff makes repeated requests to Batra to take legal action. However, Batra does not take any legal action to help the Huff family and states that it is a Michigan issue. However, due to Huff's personal experience working in the courts, Huff is aware that federal action could be taken, and Ravi takes no actions on Huff's behalf.

On Saturday, October 30, 2021, Huff first detects that he is being followed by vehicles. Huff was trained in counter surveillance detection in the US Army and while as part of routine training as a senior member of technical staff for top secret security holders at the country's nuclear weapons laboratory, Sandia National Laboratories. He initiates movements according to training to detect if he is being followed, and he observes a sophisticated tailing operation comprised of multiple vehicles while driving between his house in Champion, MI to the city of Marquette, MI. Also, while shopping at Dunham's Sports purchasing firearms and security related equipment like cameras, Huff observes that there are people following him in the store who are watching his behavior and not purchasing anything. They abruptly leave the store. On the return trip home, the Huffs observe a drone flying over the city of Ishpeming, MI at a high altitude of seven thousand feet, and the drone travels to Huff's property in front of their vehicle so that the occupants can see the drone while driving. The drone then hovers over their property at the southeast corner and descends to about three thousand feet of altitude.

On Saturday, October 30, 2021, Huff contacts a licensed concealed carry pistol instructor near Ontonagon, MI, and explains the strange situation and his personal backstory and requests a personal class on an emergency basis on Sunday, October 31. The instructor agrees to provide the instruction, and Huff completes the class and receives his concealed carry class completion certificate.

On October 31, 2021, Huff reaches out to his professional network to get in contact with the Federal Bureau of Investigation (FBI). Huff's professional colleague from PepsiCo, Jason Bashura, a national security expert and former FDA weapons of mass destruction (WMD) expert in using the food system as a mass casualty delivery vehicle, re-introduces Huff to Agent Jody Stanley, the FBI's WMD coordinator in the State of Michigan. Agent Stanley and Huff had met before at other professional meetings.

Since October 29, 2021, the drones have followed Huff and his family as they drive in their motor vehicles and move about their property. The drones continue to fly over the residence each day and night. The two types of drones are observed as rotary wing with four to five feet in total size, electric hybrid powered, and large and thin fixed wing drones with a twelve foot wingspan. White colored drones are flown during the day and black colored drones are flown at night. Photos and video footage of the drones have been taken by the Huffs (the plaintiffs in Renz's TRO) and others.

During this time, and continuing until the time of publication, Huff and his family are followed by motor vehicles, some of which a private investigator hired by Huff confirmed as having license plates registered to the Michigan Secretary of State. The investigator indicated that undercover vehicles are registered in this manner.

On October 31, 2021, David Lopez from LinkedIn suggests to Huff that he go on the run, specifically to a big city. Huff is highly suspect of this advice and continues to do his work and stays at his home with his family.

On October 31, 2021, to the present day, the house has been targeted by a long-range acoustic device, typically in the evenings and afternoons, which produces loud vibrating noise inside the house and can be heard emanating from a location to the southwest by the plaintiffs.

On November 1, 2021, Huff drives to the Marquette Sheriff's Department and requests to meet with the sheriff. Huff meets with Sheriff Zyburt and his under deputy and explains to him the events over the past week, including the drones, and Huff requests that the sheriff process his concealed carry permit quickly. Interestingly, the sheriff tells Huff that a former CIA employee living in Marquette County has reported drones at his property to the sheriff. The sheriff deems the threat to Huff as credible and processes the permit within two days.

On November 4, 2021, the Marquette County Sheriff's Department confirms seeing a drone in a video. This video is later deleted by unknown hackers.

On November 5, 2021, Andrew and Emily Huff go to the local FBI office in Marquette, MI, and the Huffs report the crimes being committed and the strange threats, including the constant drone surveillance and vehicle surveillance. Andrew Huff also provides a very brief and simplified explanation of his whistleblower complaint to Agent John Fortunato. Agent Fortunato provides Andrew with his contact information. The strange threats reported to Agent Fortunato increase after Huff meets with Fortunato; however, the threats slightly change and become exactly as Huff reported to Fortunato.

On November 6, 2021, a video of a drone at the Huff residence is sent to Agent Stanley. Stanley confirms the drone and claims that she reported the drone to the Federal Aviation Administration (FAA) and to the local FBI office in Marquette, MI.

On November 6, 2021, the Huffs' adjacent neighbor "Mike O" confirms that he has seen drones and unknown people in the area around his property. The Marquette County Sheriff's Department also confirms seeing the drones on this date. Huff also uses their Starlink Satellite system's line of sight tool to detect if there are objects in the sky blocking the system's view of the sky and several objects are detected blocking the Starlink Satellite's "vision" in the middle of the sky. This information corroborates the drone visual identification by multiple parties and the location in the sky where the drones are observed.

Between October 31 and November 7, 2021, Andrew and Emily Huff communicate with FBI Agent Jody Stanley and FBI Agent John Fortunato to report the surveillance. Agent John Fortunato refuses to investigate the matter, and in fact gives misleading and false registration information about the vehicles which have been following the plaintiffs, saying that vehicles of a different make and model than the makes and models observed are registered to the plate numbers provided by the plaintiffs. The only advice given by Agent Fortunato is a request that the plaintiffs not post anything about the matter or the FBI's involvement on social media.

Huff requests that Agent Stanley report the drone activity to the FAA. The plaintiffs believe that Agent Stanley has either not reported the drone activity to the FAA or has detracted and discredited the plaintiffs in any report she may have made to the FAA.

On November 7, 2021, Huff reports the surveillance to Sandia National Laboratories and the Department of Energy as required by law due to being a former top secret security clearance holder. He speaks with Christian Gaxiola at Sandia National Laboratories. Gaxiola tells Huff that he contacted the FBI, and the FBI told Gaxiola that they are investigating the case.

On November 7, 2021, the plaintiffs continue to photograph the drones and discover that trail cameras on the driveway have been tampered with and photos saved on the cameras have been deleted. Deleted trail cam photos continue to occur on these devices throughout December until finally they are stolen. One of the cameras becomes unusable.

On Sunday November 7, 2021, Huff plans an ATV route with his wife Emily to collect intelligence on the scope and extent to which Huff is being followed, and to help determine if tracking devices were present on their 2017 Dodge Ram or 2018 GMC Yukon XL. He plans verbally via cell phone with Emily a route from their property that is approximately 7.58 miles on an ATV headed east a few hours before sunset. He plans to intentionally stop and camouflage his ATV out of plain view from the ATV trail at a hilltop facing his property north on a long straightaway. The conditions are warm and clear. He then proceeds to the hilltop and sets up an observation point to see where the drones may be launched from. While waiting on the hilltop, about ten minutes after his arrival, an ATV approaches his position on the middle of the straightaway, and then suddenly slows down where Huff is camouflaged on the hilltop. Huff observes the occupants looking at his position. Thinking that it was highly strange, and concerned for his safety, Huff decides to leave his observation post and head further east down the trail. He stops at the intersection of a park and county road PI. Huff decides to wait for dark to return home on the same route, and the ATV that previously slowed down on the straightaway approaches Huff's parked position from the west and continues down the trail. About thirty minutes later, the sun sets and Huff begins to travel to his house with no lights turned on his ATV. As Huff approaches Carp Creek, there is a large opening without trees, and there is a man, 6'2", about 225 pounds with blonde hair, burning a campfire next to the ATV trail. Huff makes eye contact with the man. As Huff passes the man, he notices two drones hovering near this man's position at about three to four thousand feet in altitude. The drones and the man proceed to follow him in his two-passenger side-by-side. Huff takes a faster alternate route to Highway 41. Two cars begin to follow Huff and have to travel slower than the surrounding traffic to follow Huff's ATV which cannot travel faster than about 50 mph. Typical traffic speeds are 60–65 mph on Highway 41. The two vehicles follow him to county road CKC at the corner of Huff's property and abruptly turn north up CKC, adjacent to Huff's property in the east. Huff did not bring his cell phone with him. Based on this event, Huff begins to think that his cell phone was hacked and that his conversation with Emily was eavesdropped on.

From November 9 to December 6, 2021, Huff hears what sounds like all-terrain vehicles leaving residences in the middle of the day every time he leaves the house to go hunting with his dogs on their 171-acre property. A hunting walk can easily take thirty to forty-five minutes in each direction through the woods on Huff's property. The ATVs sound like they are leaving a property to the north and east of the Huff property and traveling south down county road CKC, then west down Highway 41 for about a half a mile, then turning north onto his driveway, and then driving .8 miles up his driveway to his house. The ATV would idle for several minutes at what sounded like his house and then would drive back on the same route once he could hear the ATV leaving his property. Later, ATV tracks are observed turning onto his gravel driveway from Highway 41, as well as on his driveway, with tracks going to the house near the main entrance. There is only one neighbor near the entrance to the driveway that works days, and there are no roads frequently trafficked by ATVs in the vicinity of the plaintiffs' driveway. The land surrounding the plaintiffs' property is not navigable by any type of vehicle due to water, swamp, bogs, and dense forest cover. This activity continues until the first snowfall in December 2021.

Anthony Ramos, former EcoHealth Alliance communications director and friend of the Huffs, refers Andrew Huff to journalist Katherine Eban, who claims she wrote for *Vanity Fair* magazine. Later, Huff discovered Katherine had published a favorable article for EcoHealth Alliance in the past and was not likely acting in good faith. Eban later publishes an article based on Huff's documents and his interviews with her, and his name is not mentioned.

On November 10, 2021, Huff is followed several times by motor vehicles and individuals while on a business trip for MTRX INC and while at the airport in San Antonio. Also during the work trip to San Antonio, Texas, Huff noticed people following him to and from the hotel as he visited friends at a local bar and on several trips to a convenience store. The vehicle was a bright red small compact four-door sedan, from an Asian manufacturer—Toyota, Honda, or similar. This vehicle and a new white Ford Explorer follow him to and from a corporate-sponsored event for MTRX INC. The venue was thirty miles away from the hotel and required personal or bus transportation to and from the event site. Upon completion of his MTRX duties in San Antonio, Texas, Huff has to drive from San Antonio Austin, TX to catch return flights to Marquette, MI.

Upon leaving the hotel to drive to the Austin airport, Huff suspects that he is being followed by the same white Ford Explorer as previously identified. On the freeway headed north, he abruptly pulls over and stops on the interstate after a blind curve entering a straightaway. Not expecting this maneuver, an older, seemingly tall white male is observed in a recent model white Ford Explorer that was at least a quarter of a mile behind Huff's vehicle, and the vehicle engages its brakes and slows down slower than traffic, and then rapidly accelerates above the rate of traffic. After a minute, Huff signals and re-enters traffic at a slower rate of travel, 50 mph. As Huff approaches a forced lane split to continue to Austin, the Ford Explorer is seen waiting on the shoulder approximately fifty feet before the freeway interchange lane split. As Huff approaches the split, he puts on a turn signal indicating to take a right-hand split/fork, where the Ford Explorer re-entered traffic and took the split north toward Austin. Then Huff changes his direction of travel and takes the opposite lane split and heads the opposite direction. Then, Huff drives a circuitous route which amounts to roughly a six-mile circle through residential neighborhoods. Huff reenters the freeway heading north using the exact same freeway entrance and re-approaches the same split on the freeway where he previously separated from the white Ford Explorer. Upon approaching the split, Huff observes the same white Ford Explorer parked near the split. Presumably the driver and vehicle exited the freeway and somehow anticipated his direction of travel or location. Once again, as Huff approaches the split, the vehicle re-enters traffic and goes to exit north to Austin, and Huff once again takes the opposite split and drives a different route to Austin. Huff suspects that his phone is being tracked and turns it off.

Upon arriving at the airport, Huff drops off the rental car. While walking from the car rental drop off to the terminal, Huff notices a woman in her early thirties with a medium build, approximately 5'6", following him. She does not have any baggage, which is very strange for an airport. The woman follows Huff to the terminal, then walks over to passenger pick-up area, where a late model blue/gray Chevrolet Impala pulls up to pick the woman up. The driver is a white male in his early thirties, with dark brown or black hair and a goatee. Huff takes several photographs of this woman, which are deleted from his phone by hackers. Later, Huff identifies this woman three additional times in the West Ishpeming, MI area. Once at the BP in West Ishpeming while driving a light, newer blue Chevrolet Tahoe with four-wheel drive on two separate occasions. At the BP, she watches Huff and does not purchase anything, and leaves when Huff does. There are other suspicious persons in both airports and while traveling. These people are all oddly waiting for Huff (Huff deliberately takes long, drawn-out pauses at odd locations as part of his counter surveillance training) or following him around the airport. Huff takes pictures of the woman in both Texas and Michigan, and the photos are later deleted from his Apple iCloud account and hacked iPhone.

On November 11, 2021, Huff observes an attempt to delete files from his personal computer. He observes the hackers attempting to delete MTRX records, personal records, and EcoHealth Alliance records. The hackers successfully delete the unencrypted partition in the hard drive; however, the encrypted backups are not deleted. This hack was reported to Agents Stanley and Fortunato, who refuse to investigate the theft of data that poses a national security risk. On this date, all Huff's accounts are hacked, both personal and business. He is unable to access accounts or the internet on his phone or computer. The system remains down until January 10, 2022. Huff abandons the first security company who comes to install a new router, as he has reason to believe either the router or the company are compromised, when the technician was unable to change the administrative password on the newly installed router. Huff uses a different local vendor and installs a secure router and regains internet access.

On November 11, 2021, all networks and devices in the Huff household are hacked. The Arlo Security system will not respond, networking routers and network hardware does not work, information is missing from connected devices and computers, and Huff's phone is not responding to touch input. During the discovery of the hacked network, an erroneous network appears named "Starlink–PIA–We got your back" and is observed to be functioning in the house. Huff's nearest neighbor is a mile away.

On November 12, 2021, Andrew Huff drives to the Marquette, MI FBI office and meets with an agent, and offers to turn over all his hacked equipment to the FBI. The FBI refuses to accept and analyze his hacked equipment and tells Huff to hold on to it. The agent working the desk tells Huff to "look around in the trees near his house for surveillance equipment." Huff tells the agent that his suggestion is absurd since there would be no way to power the devices. The agent stands up and leaves.

On November 12, 2021, the Huffs travel to Minneapolis to visit family. The trip was not discussed out loud and all devices were put in a Faraday cage (a device that blocks radio frequencies and electronic transmission of data) before and during the journey.

On November 13, 2021, at the Minneapolis VFW, two men are observed by Emily Huff's cousin, Nicholas Rigert, taking photographs of Huff. Huff's relatives photograph the men.

On November 14, 2021, the Huff family and extended family go to the Mall of America where Huff decides to run a counter surveillance route for 4 hours at the mall and detects that he is being followed by a team of 10 people. These people are not buying anything and after 3 hours the same tails begin to re-appear from Huff's first arrival at the mall. Huff takes photographs of these people. Huff later sends these photographs to FBI Agent Stanley where she claims that she is uploading them into some system for tracking or analysis. Later, hackers delete these photos from Huff's other devices and cloud storage; however, these photographs were backed up locally and relatives of the plaintiffs' maintained for safe keeping.

On November 15, 2021, on the way home from Minneapolis, MN, Huff stops to replace the SIM card in his phone as he suspects it has been hacked. After a factory reset, the WickerMe app, recommended by David Lopez from LinkedIn, reinstalls. He decides to detour to the Apple Store in Milwaukee, WI, where the phone is replaced under warranty. While driving home, the iPhone becomes burning hot. This is a pattern that Huff recognizes from other times his phones were hacked. He would obtain a new phone and, shortly after purchasing it, it would become burning hot, probably due to brute force hacking. As Huff drives back from Milwaukee, he observes two drones up in the air, and the drones follow him all the way from the north side of Milwaukee to the Wisconsin and Michigan border 140 miles away.

On November 16, 2021, Huff attempts to contact Michigan cyber command to have his hacked devices investigated. Huff leaves numerous messages with the cyber command and does not receive responses. Huff later determines that the phone calls were somehow being blocked, since when he first activates a new cell phone, a flurry of missed messages and voicemails from previous weeks suddenly appear. When Huff attempts to access his voicemail, his voicemail password will not work. This occurs to at least four brand new iPhones that he obtained due to hacks over the course of four months.

On November 22, 2021, the Huffs and their extended family go to a new Mexican restaurant in Marquette, MI. While eating dinner, a man walks in who Huff identifies as the man on the trail on November 7, 2021. Huff confronts the man and calls 911 to request that the City of Marquette Police Department identify the man. The man comes outside and states that he is from Louisiana and works in the oil and gas industry, but is in Marquette for work. None of this makes sense to Huff. However, the identity and work history of the man should be verified.

Huff has repeatedly reported drone activity to Michael Matthews of the FAA, as well as the FBI. The drones are at an altitude of three thousand feet or higher, which subjects them to FAA regulation.

Huff sends registered letters to numerous federal and Michigan state government officials requesting help and stating that his constitutional rights are being violated. He receives one boilerplate response from FBI Director Christopher Wray advising him to contact his local FBI office.

On December 10, 2021, Huff observes an ATV enter his property. His dogs begin to bark in alarm, and a person enters his barn an eighth of a mile from the house. The dogs chase the person off the property. This occurs before a heavy and significant snowfall.

On December 11, 2021, Huff discovers that the key to his industrial Bercomac snow removal attachment has been removed, and the Huffs must travel thirty-five miles to borrow a replacement key. Once the key is obtained, the Bercomac is not running properly.

On December 11, 2021, the plaintiffs discover that both of their vehicles have been hacked. While driving down rural roads, both vehicles, a 2017 Dodge Ram Rebel and a 2018 GMC Yukon XL, suddenly steer left and right causing the vehicles to go off the road. The vehicles are also shifting strangely and multiple electronic systems not working properly in both vehicles. The vehicles smell like something is hot or burning. The recently installed snow tires have sudden leaks. The tires did not have any leaks when they were removed from the vehicle the previous spring, and they were stored in a barn where people were observed trespassing and messing with the plaintiffs' equipment.

On December 14, 2021, a US Coast Guard helicopter is observed hovering and circling over Huff's property at roughly 8:30 a.m. Huff takes a picture of this helicopter and sends it to several friends and to a private investigator who positively identifies the US Coast Guard helicopter over Huff's house.

Later on December 14, 2021, Huff's business partner at MTRX INC and former engineer at EcoHealth Alliance, Brock Arnold, reports to Huff that US government Blackhawk helicopters approached his house, hovered over his house in Malone, NY for an extended period of time, and then changed directions. His brother Robert Arnold is a witness to this event. This indicates that Arnold's house was a designated waypoint on a flight path. The Arnolds report that it happened around the same time that the Coast Guard helicopter was hovered over the Huff residence. Andrew Duso, a MTRX Inc. executive, reports that people are following him in his community of Malone, NY.

On the morning of December 15, 2021, Huff takes the dogs for a walk and observes drones flying and flashing white lasers near his dogs to agitate them. After this walk, there is a low cloud deck and Huff hears what sounds like an Air Force C-130 circling around his property, low but slightly above the clouds. He notices that all the home's windows are emitting a loud and low-frequency vibrating noise. As Huff walks closer to the windows, he realizes that his house is being hit with a device known as a Long-Range Acoustic Device or an Active Denial System as his ears begin to ring and his skin tingles.

Mid-day on December 15, 2021, a drone is hovering in front of Huff at fifty feet. Huff observes the drone and can tell that it is an advanced electric gas hybrid model used by the US government with highly sophisticated optical systems mounted underneath its fuselage. This drone is a quadcopter, with a white protective shell, is highly smooth and aerodynamic and is approximately four to five feet in diameter.

On the night of December 17, 2021, three or more drones hovering on the edge of their property at three thousand feet of altitude or higher flash red, yellow, and white lights in the Huffs' bedroom window while they both attempt to sleep. From this point forward, the drones routinely flash lasers in the windows at their two dogs to agitate them.

On December 17, 2021, a man walks into the laundromat in Ishpeming, MI, and tells Huff Huff's own life story. By this point, Huff has determined that someone has launched a psychological operation against him, and he laughs it off, takes a video of the man, and leaves the laundromat.

On December 17, 2021, the plaintiffs take the 2018 GMC Yukon XL to have its systems examined. Huff receives a detailed report from the technician, which contains a long list of strange diagnostic trouble codes. The technician resets the modules and some things return to normal. The vehicle drivetrain (the group of parts that interact with the engine to move the wheels and various parts of the vehicle to thrust it into motion) continues to smell hot, and the gas mileage decreases by almost 10 mpg on average. This vehicle has not been the same since it was hacked.

On December 17, 2021, while at the shop having the GMC car serviced, Huff observes a man and a woman in the customer waiting area. Neither person had vehicles in the shop for service. Huff observes the man appearing to run code scripts as he watches his screen from a few feet away. He was at the dealership sitting in a cubicle when Huff arrived. Huff, knowing that his devices are being eavesdropped on, clearly jokes with his wife about arresting the man at gun point via text message. Then, a woman enters the dealership customer waiting area, sits, and directly watches Huff, presumably to provide security for the hacker. Suspecting that his phone has been hacked, Huff makes jokes about the woman's appearance to his wife via text message, and the woman's facial expression becomes bitter while staring at Huff. Huff moves to a seat out of her direct view, and then the woman repositions herself to maintain visual of Huff. When the technician comes out to speak to Huff about his vehicle, the man and woman suddenly leave the customer waiting area and dealership together. Huff reasonably believes that the man was trying to hack the computer system at the dealership to delete the evidence obtained from the hacks before the technician discovered them.

On or about January 1, 2022, the Huff residence is forcefully entered via the garage door, which is forced up. Also on this night, Huff goes to a small local bar in Republic, MI. While enjoying a cocktail, two people from out of town walk into the bar, presumably a couple. The woman is very attractive and proclaims to be Ukrainian, which Huff believes to be accurate, and has a much older husband. The couple say that they are staying at an Airbnb in Republic to go downhill skiing. Huff immediately believes that these people are federal agents of some type. Huff explains to the couple what has been happening to his family for the past few months and tells them that he suspects that they are agents. The people are nice and agree to be patted down by Huff for weapons. As the bar closes, they extend an invite for Huff to go back to their Airbnb for additional drinks, and the man suggests that Huff have sex with his wife. Huff agrees to go to the Airbnb on the condition that he will always have his firearm on him. Huff and the couple make small talk, and he learns that the man was recently released from prison for methamphetamine distribution when he begins to talk about dealing drugs. None of their story makes sense and Huff believes that this is a strange diversion tactic or an FBI entrapment scheme. Huff stays until the morning and leaves when he believes that he is safe to drive a vehicle after consuming alcohol.

The break-in is discovered on January 3, 2022. Huff finds damage to numerous items external and internal to the house. Non-Wi-Fi capable devices appear to be operated by Huff's hacked phone, and household appliances and the wood boiler appear to be operated and controlled by cell phone. This becomes obvious when Huff turns off his cell phone and all his household appliances that were acting oddly return to normal. There is no wireless network at the house due to the hacks on November 11. The closest house is a mile away. It appears that someone attempted to bring down all the systems in the house. He discovers that wall outlets and other receptacles have been tampered with, and control and/or recording devices were installed. Later, it's discovered that a corrosive was dumped into a new wood boiler, the electrical generator was tampered with, components of the power inverter were tampered with, solvents were dumped into engines, storage batteries had been tampered with, and a hot tub no longer blew bubbles, among many other damaged items. The plaintiffs have these devices in their possession and they file a police report. The plaintiffs have thus far obtained reports from experts that indicate that a state-sponsored actor is responsible for Huff's hacked computer.

On January 3, 2022, Emily Huff is attempting to leave for a work trip. As she is leaving, the 2017 Dodge Ram Rebel becomes stuck on a mildly slippery surface, which normally would not be an issue, even in two-wheel drive. Emily Huff attempts to place the vehicle into four-wheel drive. When she places the vehicle into the drive position, the vehicle reverses and accelerates, while still in the drive position. Andrew yells at Emily, thinking that she made an error, until he investigates the vehicle himself and observes the vehicle's transmission in the drive position. The electrical systems begin to act crazy. Then a loud cracking noise is heard, and the vehicle becomes inoperable. The vehicle has remained parked due to significant safety concerns.

In the early morning hours of January 4, 2022, trespassers are observed on the property shining lasers in the windows from the ground onto the ceiling of their second-floor loft at 4:30 a.m. while Huff was attempting to sleep. The laser beam light stream and angle indicates that the trespassers are located near the driveway. Huff calls the police and observes the trespassers running down the driveway. The driveway is packed with snow and ice, so no footprints are left. Huff also reports the break-in and reports the tampered vehicle, generator, wood boiler, and missing hard drives and thumb drives.

Since early January 2022, community business owners, community leaders, and neighbors report that a late model Chevy Tahoe and a late model Ford Expedition in the Marquette area has been driving out to the area of the plaintiffs' property, and a late model Ford Expedition. Both reportedly are equipped with advanced electronic equipment and strange antennas. The plaintiffs observe these two vehicles in the vicinity of their property on several occasions. Community business owners reported that during this time people rented storage lockers for advanced electronic equipment.

On January 8, 2022, Andrew and Emily Huff go on a date in Marquette, MI. While having beers at a local brewery and playing darts, a heavy-set man, about 5'11" with a crew cut, red hair, a military or law enforcement style mustache, and black glasses, sits at a table facing the Huff's and pulls out a notepad. Then the man begins to take notes and closely watches both Emily and Andrew in the otherwise empty large establishment. Andrew Huff confronts the man and asks him if he is a federal agent, and the man states, "Yeah, I am CIA." Huff suspects that he is FBI, or a federal law enforcement agent based on his appearance and behavior.

On January 9, 2022, MTRX Inc. executive Andrew Duso reports that his entire network has been hacked, which forces Huff's company to cancel a partnership and investment meeting on January 9 with OSF Healthcare. This results in a $2 million dollar loss to MTRX Inc.

On January 17, 2022, Huff communicates with Senator Gary Peters's office. All seems to be going well. Then a staffer from Peters's office apparently threatens a colleague of Huff's, Lindsay Kate, with providing false COVID information and points to the DHS threat alert related to COVID misinformation and says that Huff had better have an attorney on the call.

On January 20, 2022, Huff speaks with Col. Derek Harvey, lead House Intelligence investigator, Dan Berlin, ex CIA officer, Staff Patrick Davis from the House Intel Committee, Congressman Brad Westrup, who was serving on the House Intelligence committee, and staff from Congressman Jack Bergman's Office. Interestingly, Harvey, Davis, and Berlin all tell Huff that they have heard stories like Huff's before, some even more bizarre than Huff's, and that the US government has engaged in these types of activities before. Huff also speaks with Patrick Hartobey from the US Senate Minority Party Committee on Investigations and provides both the Senate and House committees copies of all of Huff's files to support Huff's whistleblower claims.

On February 11, 2022, Huff files formal whistleblower complaints with Office of the Inspector General—USAID, HHS, USDA, DHS, OIG IC and with Michigan Senators Wicker, Burr, and Peters. On that date, he returns home to find his mailbox lying on the ground in the snow. It does not appear to have been struck by a snowplow. The mailbox is perfectly intact, with no strike marks, lying perfectly flat on the ground. There are no tumble marks in the snow. He files a police report with the Marquette County Sheriff's Department. He places the mailbox upright on the snow to continue to receive delivery.

On March 4, 2022, Agent Jody Stanley randomly contacts Huff and heavily pressures him to speak with her. She offers to come to his residence from Detroit, even though there is a local FBI office only twenty-five miles away where Agent John Fortunato is located. She wants to speak with him "about the things that you have been saying." She then offers Huff a "quid pro quo." She will only investigate the crimes against the plaintiffs if Huff speaks with her about the "things that you have been saying." Huff refuses the interview. Agent Stanley also claims to have attempted to call both Renz and Huff. Neither one has any call history from Stanley. Huff and Renz wonder if Agent Stanley is being truthful.

On March 5, 2022, Huff retrieves the mail from the mailbox and discovers that someone opened the mailbox door and fired a bullet through the back side of the mailbox while it was lying on the snowbank. Huff has the mailbox and photos of the incident.

On March 5, 2022, Huff determines that someone has launched a Denial-of-Service attack on Huff's network, as connections to websites and internet traffic are being blocked as he attempts to report the intentional mailbox damage, a federal crime, to the USPS. He attempts to submit the report twenty-five times, and all are blocked. He then reconfigures the network to stop the Denial-of-Service attack and is able to submit the report. Three attempts are made, and confirmations of the reports are received.

On March 5, 2022, the Huffs discover that their internet traffic is being strangely routed. This results in the Huff's not being able to access accounts for work and other strange problems impacting communications, work, and MTRX Inc. The persistent nature of hacks and disruption of communications has resulted in millions of dollars of business and lost market access for MTRX Inc.

On Wednesday, March 9, 2022, volunteers working with Renz Law report that Huff's messages are not arriving in a timely manner. This has resulted in lost productivity at Renz Law and indicates that the actors are interfering with Huff's claims against the government. MTRX executives also report attempting to call Huff. Huff has no record of missed calls from the executives on his devices. Calls are being blocked on Huff's phone.

Between March 9 and May 10, 2022, the drone flights over the plaintiffs' property, surveillance, Denial of Service network attacks, network hacks, internet traffic re-routing, and other harassment, tortious, and illegal activity by the defendants continue.

On May 11, 2022, Huff identifies two men that followed him on three separate occasions. This time the men are driving a new White Buick Enclave rental. The men are white and in their mid-twenties. Male 1 is 5'11" and approximately 180 pounds, with pale skin, dark brown eyes, a scruffy and patchy red-ish brown beard two inches in length (or shorter on different sightings), always wearing baseball caps, black or dark clothing like tee-shirts with a sweatshirt, and carries a medium- to large-sized black backpack with graphic patterns. Male 2 is 5'10" and approximately 170 pounds, with fair skin, shaved or short scruffy beard, dark brown hair, brown or brown/green eyes, and is always dressed in black or dark clothing like tee-shirts with a sweatshirt, and also carries a medium- to large-sized black backpack. These men were also present on a Delta flight from Minneapolis, MN, to Marquette, MI (Delta flight number DL 4258), and were sitting in Row 1, Seats C and D.

On May 12, 2022, Huff decides to check the trail cameras' SD cards aimed along his half mile long driveway. Huff and his wife leave their property at least two times per day and receive deliveries at the house. He discovers that the SD cards are all missing photographs on the same day: May

10, 2022. After making this discovery, he decides to check the heavily camouflaged trail cameras throughout the woods surrounding his home and property. On the trail cameras' SD cards, the cameras recorded and took photographs of drones hovering around his property on several different days, trespassers on side-by-side ATVs driving on a trail two hundred feet south of his home, and a large drone that crashed near a tree line with an unobstructed view into his home. This occurred on May 10 in the trail camera photos. After making the disturbing discovery, Huff calls his attorney Tom Renz to communicate with him about what he found. After calling Tom, two helicopters arrive at his home and begin to circle the property in proximity to Huff's location. Huff's neighbor Chad Vourien is also present as he is helping Huff reposition trail cameras. Additionally, at least two large quadcopter drones, estimated to be four to six feet in diameter, begin to maneuver around Huff's and Vourien's positions in his backyard. While standing on a ladder, repositioning the cameras, Huff is level with his main floor windows, and he observes a person enter his house and attempt to destroy the images that he just copied from his trail camera to his computer. The person was 5'8" with a thin build, light skin, black hair, potentially of Asian descent, and either a small-build man or a woman. Huff screams to get his wife's attention, draws his firearm, and runs down the ladder toward the house. When Huff screams, the burglar exits the front door of the house and runs north into the woods. Huff stops pursuit to check on his house and the safety of his family. The sheriff's department is contacted, and Attorneys Phyllis Crespo and Jeff Stelzer from Renz Law speak with the captain and deputy of the Marquette County Sheriff's Department the next day with Huff via a video call.

On May 13, 2022, the Huffs decide to check their security camera hard drives to make video copies of the intruders for the Sheriff's Department. Upon checking the Network Video Recorder (NVR) video storage interface and associated drives, the Huffs discover that the time and date are incorrect on the security camera recorder. They also discover that videos from the security cameras facing the front door have been tampered with to display a loop from a previous day or have been deleted. While watching the NVR system's playback footage for all the cameras during the attempted burglary period, they see a short clip of the intruder on the west side of their house, getting ready to approach the house. While discussing the discovery, the video is deleted. It is important to note that the NVR system cannot communicate with the internet since the NVR security recording system has never been set up to be connect to a network. They also discover that the NVR system is behaving strangely on all IP cameras' recorded video, as the recording time steps are off between the cameras and the NVR system. They surmise that when the large drone crashed, the house was broken into to destroy the evidence captured by the NVR camera recording system, since the cameras facing the backyard on May 10 are missing from the recording system. Huff decides to inspect the house and discovers a few electrical devices have been tampered with or moved but are still in good working order. The Huffs contact their security company Range Telecommunications and request emergency service.

On May 14, 2022, several small drones are discovered in the house and two are captured. Some of the drones are the size of mosquitos and glow green from what appears to be a tiny fiber optic cable that is about the width of a hair. They make noise like a mosquito, but fly and maneuver like an airplane at a speed much faster than that of a mosquito. That same day another drone disguised as an insect (about the size of a fifty-cent piece) is discovered. It displays similar flight characteristics to the other tiny drones and flies like an airplane at a very high rate of speed. It has red fiber optic cables lining its body. When they first attempt to crush the drones with magazine or other hard objects, they are able to regain and resume flight after being struck. The Huffs attempt to capture all the mini drones without luck due to their speed. As the Huffs perceive that they are there merely to pester and annoy them, they eventually use an electric flyswatter that administers an electric shock and successfully disables the tiny drones.

On May 15, 2022, technicians from Range Telecommunications arrive and perform an inspection of the intrusion detection system, and strange alarm activation and deactivation times are discovered along with numerous attempts to connect. It appears that the intrusion detection system's LTE signal was jammed during the suspected break in on May 10, 2022. The technicians document the irregularities of the NVR video recording system, namely, that video loops had somehow been spliced into the camera's feeds during the break-ins and that the video recorded on the loops inserted into the DVR system did not match the environmental conditions that were present at the day and times where the splicing occurred in the recorded video feed, that the times on the IP cameras were not synced with the NVR system clock, and a video of the technicians narrating and describing the problems while the camera feed was projected on a large screen was recorded. The technicians take logs of the devices for deeper analysis, and no problems are found in the intrusion detection system or the NVR system.

On May 16, 2022, Andrew Huff's Apple Mac Book Air laptop goes missing and is presumed to be stolen from their house. While sitting on the couch in the afternoon, Huff observes a trespasser outside somewhat hiding in a concealed position behind fallen logs and brush carrying a firearm who is preparing to launch a drone that is about one foot in diameter. Andrew tells Emily Huff to call the Sheriff's Department to report the active crime. Huff decides to use an infantry tactic that he learned in the Army to push the person into a swamp where there is no way to easily escape before leaving the house. Huff exits the house and runs rapidly west with a handgun and shotgun; the criminal is positioned south and west of the house. He runs past the criminal's flank and then immediately pivots ninety degrees and charges at the position where the criminal was hiding and sees brush breaking ahead of him heading south into a swamp. Huff pursues the man into the swamp where the criminal displays his firearm in the distance, and Huff immediately engages him with the shotgun, forcing the man into deep mud and mangrove-like brush. Huff pauses to listen for movement and reengages the target. When 60 percent of his ammo has been exhausted, Huff maneuvers to a better covered and concealed position. Huff hears movement in the swamp again and briefly engages the target. During the next pause, the criminal stuck in the swamp deploys a tactical rescue beacon which launches into the air at a height of roughly sixty feet and makes a very loud, low-pitched electronic "whooping noise" followed by a high pitched chirp. Huff hears two vehicles drive along the west side of his property, and he believes that they dismount their vehicles near a park which is two miles from his position to the southwest. Gunfire then begins coming in Huff's general direction from the southwest and he returns fire. Huff exhausts his ammunition and then runs back to his house to get more ammunition and check in with Emily. Huff has Emily call the Sheriff's Department again and leaves the house to return to his position. Gunfire erupts again from the southwest, and it sounds like a large caliber handgun. Huff determines that the person is using it as a distraction to cover the sound of the trapped criminal maneuvering east toward their driveway, the only potential exit point from their land. Huff sprints to the choke point and waits while he can hear the criminal moving through the swamp. Huff opens fire into the direction of the sound and the movement stops. While this is occurring, a late model silver GMC half ton with loud exhaust and tinted windows drives past the entrance to Huff's driveway no less than four times. Huff ascertains this is the criminal's escape vehicle. While waiting for the criminal in the swamp to move, Huff hears two to three loud sirens approaching his house. Huff places his firearms on the ground, expecting the sheriffs to turn up the driveway. Instead, the sheriffs drive past Huff's driveway heading east at a very high rate of speed. Huff decides to lecture the man in the swamp about engaging in illegal actives on behalf of the US government and returns to his house. Huff calls the Sheriff's Department to determine where and when the sheriffs will arrive. The dispatcher tells Huff that his house is the next on the list, and that there was a car accident on Highway 41 near Co. Road 478, a few miles east of Huff's property, that the sheriffs were diverted to. Huff requests a copy of the police report from the dispatcher, and she states that he can pick it up tomorrow. The next day Huff attempts to retrieve the police report, and the Marquette County Sheriff's Department states that they cannot give it to him even though the accident occurred on public roads.

On Friday May 20, 2022, the networking equipment at the Huff's residence has stopped functioning. The Huffs need a break from the harassment, so the Huffs decide to take a vacation to visit family in Massachusetts. The Huffs decide to drive to Massachusetts and drop off their GMC Yukon XL, which has an upper control arm on the driver's side which has either been cracked or slightly cut with a saw. While dropping off the Yukon XL, there are two people watching the Huffs while walking around erratically in the Fox Naguanee GMC parking lot. They are both white, approximately fifty years of age. One is a man about 6'1" and 195 pounds with four-inch-long tapered brown hair, and the other is a woman about 5'7" and 155 pounds with long gray and brown curly hair hanging mid-chest, driving a mid-2000s black Ford F150. Huff initiates contact with woman and describes her voice as sharp with a medium tone. The Huffs, who are quite used to the harassment, make light of the situation and begin to travel in their Dodge Ram to Massachusetts. Drones being controlled by passengers in vehicles following the Huffs follow them to Indiana, and vehicle tails follow the Huffs all the way to Massachusetts. No abnormal activity is noticed in Massachusetts during the one-month long trip. While in Massachusetts, the Huffs speak with technicians at the Fox Negaunee GMC where their Yukon XL is being repaired. Quite strangely, the dealership reports that the control arm which they both observed as being cracked is fine and looks new, although the brake lines are damaged and need to be replaced.

Upon their return home to Michigan from Massachusetts on June 30, the drone harassment and hacking of the Huff's computers and phones continues.

On September 1, 2022 the Huffs drive from their home to Chicago, Illinois to attend Arc Music Festival and stay at the Kimpton Gray Hotel located at 122 W Monroe St. While attending the music festival, Andrew believes that US federal law enforcement agents are following them around the festival as these suspected agents do not fit in with the crowd. They notice men following them pretending to be homosexuals and older attendees wearing high end luxury watches but wearing very inexpensive clothing, which blatantly stands out in the crowd. Some of these suspected agents repeatedly offer the Huffs what appear to be illegal drugs, which they refuse. The Huffs' friend Mr. Micah Ternet, a combat veteran friend of Andrew, also observes the strange people and their strange behaviors at Arc Music Festival on the third of September. On the night of September 3, Andrew identifies tiny hidden cameras placed in his hotel room (Room 402) which are cabled to the room above his (Room 502) and someone outside the room is flashing a laser into his room, which he observes reflecting off the window across the alley and sees it is being operated by two men in the room above his in room 502. Mr. Huff goes to confront the men in room 502, and the men do not answer the door. Andrew exits the room into the stairwell when he hears the two men exit the room and run down the hallway. Andrew, anticipating their hasty exit, is already waiting in the stairwell on the sixth floor and the men enter the fifth floor stairwell and run down the stairs heavily intoxicated. Male 1 is approximately twenty-six years of age, has blonde hair, is tan, is roughly six feet tall, is wearing dress slacks and a dress shirt, and had medium build. Male 2 is approximately twenty-eight years old, appears to be six foot two inches with fair skin, and had dark brown or black hair, with a medium to thin build, wearing dress pants and a dress shirt. The incident is reported to the Chicago police, who refuse to investigate.

On Wednesday, September 14, Dr. Huff notices that his 2018 Yukon Denali is acting strangely, and notices that the incorrect ECM (vehicle computer) is installed on his vehicle. He calls 911 to report the incident and is told that a deputy will contact him. Shortly thereafter, Dr. Huff hears a phone ringing in the bushes of the back yard and then his personal cell phone rings from this number: 989-313-7756. Andrew can hear the man talking in the bushes and plays along with the caller that identified himself as Trooper Bray from the Michigan State Police. Andrew decides to play along with the "trooper" while he can hear Trooper Bray giggling in the bushes with another man. At the end of the phone call, Andrew identifies more people trespassing on his property and calls 911 to obtain an estimated response time. Shortly thereafter Trooper Jared Wells and "Trooper Miller" arrive in a Michigan State Patrol Ford Explorer. Andrew immediately identifies Trooper Miller as the FBI agent from the Kimpton Gray Hotel that was shining a laser in his window and that ran in front of him down the stairs (phone numbers of Wells and Miller: 906-458-4752; 906-281-6217). When these men arrive, Andrew's neighbor Chad Vourien is present with a thermal optic camera along with wife Emily. Also, Attorney Tom Renz is on the telephone throughout the interaction. Andrew pretends not to recognize FBI Agent Miller throughout the interaction, where Miller proceeds to lie about who he is and provide cover for Trooper Bray who was observed and heard in the woods by Huff and observed by Vourien with his thermal optic camera. Both Trooper Wells and Agent Miller refuse to investigate or attempt to detain the trespassers observed in the woods by Huff and Vourien. Huff reports to Wells and Miller that sensitive national security data was stolen from a safe in his house, along with Agents John Fortunato and Jody Stanley. No investigation occurs.

Dr. Huff decides to wait until the next day, September 15, 2022, to call out Agent Miller for being an FBI agent with Tom Renz via text message. Since the identification of "Agent Miller," the Michigan State Police, and the FBI as being the sources of harassment, the civil rights violations have greatly decreased.

On the weekend of October 8, Huff discovers two of the IP cameras had wireless access points installed. Huff also discovers a hidden network running inside his home with three access points, which are connected to seven or eight unknown devices in or around his property.

On the weekend of October 22 the hacking was successfully traced to a Department of Defense (DoD) data center in Ohio.

Until the present day, the Huffs are continuing to experience harassment from the FBI, DoD, and Michigan State Police.

***

How much do you think that this operation cost US taxpayers?

My estimate is between $45 and $55 million. After the first month of harassment, which began in late October 2021, I started calculating what it would cost per day to have a twenty- to thirty-person team running an operation, including local travel, and equipped with the best advanced technology in the world,. At first, I was trying to determine who the opposition was, but it became obvious when the US Coast Guard hovered a helicopter over my property and house, nowhere near the coast.

What you just read has been rather hard for everyone in my life to believe, but several members of my family and neighbors, as well as numerous people in my local rural community, several members of local law enforcement, and my attorney Tom Renz, have all been witnesses to everything I wrote.

Since I was trained in how to detect and perform counter surveillance, I collected numerous license plate numbers. As my story and my whistleblower claims began to spread, I was introduced to Robert F. Kennedy Jr., who was very interested in what I had to say about the CIA in light of his family's unfortunate history dealing with them.

Sometime in February or March 2022, when I was still digging out from the perpetual break-ins, hacks, and home repairs, I asked him for help finding investigators to go after these people. So he introduced me to the famous security expert Gavin de Becker, who then referred me to trusted specialists that work with his security company, which unfortunately resulted in a $350,000 estimate that I can't afford.

What irritates me so much about all of this is that the people who have been harassing me have mostly been Americans.

Typically, the CIA or DoD do not conduct domestic operations against US civilians, and if they do, they are required to have an FBI liaison with them to conduct these illegal domestic operations.[263] [264] [265]

The aircraft have all been DoD or FBI.

FBI Agent John Fortunato, who I suspect was merely playing dumb when I asked him for help, lied to me in at least one instance with the intent of throwing me off the US government's tracks and insisted that I don't tell anyone or post on social media about their "investigation."

FBI Agent Jody Stanley, the WMD coordinator for the state of Michigan, I believe either pretended to help me or perhaps they kept her in the dark. With Tom Renz included in the conversations, I reported over fifty federal crimes committed against me, not including the numerous civil rights violations, and the FBI refused to investigate the crimes.

I sent a registered letter asking for help to FBI Director Chris Wray, and the FBI sent me a canned response telling me to speak with my local FBI Agents.

A few months later, the Department of Homeland Security (DHS) issued a threat warning about domestic extremism related to "false COVID narratives," which was conveniently released when my whistleblower claims were the talk of the town in Washington, DC, for a week, at least as I was told by DC insiders that I trust.

Shortly thereafter, when I had not given up this fight, FBI Agent Jody Stanley texted me out of blue and said that she wanted to travel from her office near Detroit to my home in a remote part of Upper Michigan to meet with me.

I started a group text message thread with Tom Renz and Agent Stanley where she stated that she wanted to come to my house, along with a few other FBI agents, to "talk about the things that I have been saying."

I was angry.

She didn't want to talk about the numerous crimes committed against me, my family, my company, or my company's employees, but she wanted to "talk about the things that I have been saying."

Then FBI Agent Stanley quite shockingly offered me a quid pro quo via text message with my attorney Tom on the thread. She said, "If you talk to us about the things that you are saying, then we will look into your claims."

So then I reminded FBI Agent and "WMD Expert" Jody Stanley that I possessed a highly sensitive dataset which I had built that contained all the DHS critical infrastructure and key resource data for food and agriculture collected from a criticality and risk assessment tool named FASCAT, to measure, compare, and rank the most attractive targets for terrorists to poison our food supply and potentially kill hundreds of thousands of people in a single attack. I collected and analyzed this data during my PhD study. In the wrong hands, this information would be a road map to kill hundreds of thousands of Americans, and it was stolen from my house. If this dataset would have been developed within the US government and not academia, then it would have received, at a minimum, a SECRET designation.

Since some of the other data probably stolen was a list of chemicals, biological agents, and other materials that were organized based on their effectiveness to successfully deploy against Americans, combined with their projected lethality when used (which was modeled and verified in actual real world food systems), not only does someone have a target list with the most vulnerable points of attack and the list of facilities that could kill the most people if attacked, but they also have a rank order list of the best agents to use to avoid law enforcement or quality control and detection.

These combined, at a minimum, would receive a top-secret designation and would be part of a special access program.

"WMD Expert" Stanley replied to the group thread, "What is FASCAT?"

I almost fell out of my chair and immediately texted my mentor who'd helped set up the biologics division of DHS during its creation, Col. John Hoffman Sr. I texted John, "What if I told you the FASCAT data and my key to it were likely stolen from my house by a state sponsored actor? You and I are the only two people that have the keys."

He replied, already aware of the illegal, cowardly, and traitorous behaviors of the US government toward me, "How do you know that we [referring to our own agents] didn't take it?"

I replied, "John, let's assume that a foreign state sponsored actor stole it."

The text bubble popped up, and disappeared, popped up, and disappeared, and then there was a long pause. He quickly typed, "That would not be good." I took a screen shot of my text exchange with Col. Hoffman, and proceeded to explain to my state's "WMD coordinator" what the FASCAT data were.

Then, I sent her the screenshot.

The thread went silent.

This text thread exchange is emblematic of the true state of national security in this country. Since September 11, billions of dollars have been spent in the name of national security, yet we have FBI WMD "experts" that are not qualified to do their jobs to protect Americans.

With all this spending on national security, we have made backward progress in national security and given up our freedoms in the process.

On a separate text thread, Tom told me to calm down, laughed, and suggested that I leave Agent Stanley alone.

The next day, Tom spoke with Agent Stanley on the telephone and then gave me a call and recommended that I not sit down for a formal interview with the FBI.

Nothing seemed right about the interaction, and she likely had made up the lie that "she was now able to travel due to the COVID travel restrictions being lifted for Federal employees."

The reality was that Agent John Fortunato and his office are only a thirty-five-minute drive from my house, and federal law enforcement personnel conducting investigations are always exempted from travel bans.

Agent Stanley would have liked to have had me believe that a COVID travel ban shut down federal law enforcement nationwide. Now I knew that the FBI had been receiving intelligence from my house and my electronic devices because I intentionally made fun of and mocked Agent Fortunato in the privacy of my own home.

The FBI thought that they had a better chance with Agent Stanley to get me to sit down for an interview, and by sending Stanley to request the interview, instead of Fortunato, I knew that the FBI was complicit in the crimes committed against me.

I told Tom that I was refusing the FBI interview, and he then let Stanley know that.

The FBI wanted me to sit down for a formal interview, just like they convinced Gen. Mike Flynn to do, so that the FBI could create a situation where I might contradict myself, resulting in what could appear as a lie.

Then, the FBI could charge me with the crime of lying to the FBI.

This is what the FBI does when they can't find anything to charge a person with.

This is the PATRIOT ACT and your taxpayer dollars at work.

I love freedom and the First Amendment.

This message is for those special people that harassed, lied, and committed crimes against my family and I:

Fuck you. I hope that the government requires you to get COVID boosted so many times that your immune system ceases to function, which then results in you contracting an opportunistic incurable multidrug resistant form of *Shigella flexneri,* and you have to spend the rest of your life stuck on a toilet rocking back and forth with diarrhea. You don't deserve the freedom and comfort in life, that relatively few people like me, have fought for and personally sacrificed for you.

The statement above is not a threat, it is hyperbole which conveys my feelings toward some people that used a professional relationship, when I was seeking help, to take advantage or harm me. This was likely orchestrated by someone higher up in the chain of command from Agent Stanley, for purely political purposes. I don't want the FBI or the government to force anything upon Agent Stanley, or anyone for that matter.

I had worked or crossed paths with Agent Stanley in the past related to my expertise in bioterrorism and food defense. My wife went to her and begged her for help while crying during the harassment, and Jody told her that she was assisting, whatever that meant.

My one-and-a-half-year-old son was anxious, frightened, and neurotic at the height of the illegal harassment.

Stanley also likely lied to an investigator in counterintelligence, Christian Gaxiola, at Sandia National Laboratories, when she told him that the FBI was investigating the crimes committed against me.

No such investigation ever took place.

These are the terrible kinds of people that work in our government and waste your hard-earned money.

I used to admire and respect all FBI, DHS, IC, DoD, and state local law enforcement personnel.

Now I believe that the FBI should be eliminated as a federal agency. I have heard some people on the news say that there "are still some good FBI agents out there." OK, please show me one so that I can change my position.

If there are good FBI agents out there, then why can't anyone point to one specifically?

The corruption and graft among federal employees in Washington, DC, has become so brazen that I am deeply concerned about the future of our country. I simply do not understand why these people can't say NO when told to engage in unethical, immoral, and criminal behaviors.

Our country used to be the leader of the free world.

The United States still might be in the lead, but our free world is disappearing before our eyes.

The only way that this will change is for everyone to stand up against them.

- Is publicly shaming the bad behavior of government employees good for our country?
- Does this book paint our federal agencies in a good light?
- Does this book make other people around the world look up to Americans?

There is one simple answer to all these questions. No.

My warning to all of you is that you could be subject to the executive branch's abuse of power for merely saying something that is unpopular, angers the elite, and is entirely true.

CHAPTER 21

Why Does the Cover-Up Continue?

In 2015, while working at EcoHealth Alliance, it became apparent to me that the business of pandemics was a massive market: vaccines, medical counter measures, medical equipment, training, research, and development.

I wrote an article that was subsequently rejected for publication in all the leading health policy and health economics journals. I couldn't figure out why it was rejected at the time.

Now, I know exactly why it was rejected.

Here is the article in its entirety.

I think that the article aged quite well, and therefore the cover-up continues.

Pandemic Winners: A Review of Global Pandemic Policy & Expenditures
ABSTRACT
Pandemics and epidemics can be economically detrimental. However, there is a paucity of research that examines what types of organizations benefit from pandemics, how these organizations benefit, and how government spending relates to global pandemic preparedness and response policies. This text explores the relationships between pandemics, expenditures, and policy to determine if government spending is improving pandemic preparedness and response to future infectious disease threats. Government expenditures and several key examples of organizations that provide infectious disease goods and services to the government are discussed. Lastly, the impacts that these expenditures may have on future infectious disease preparedness is considered.

INTRODUCTION

The economic impacts of pandemics and epidemics can be detrimental to the global economy. [266] [267] [268] [269] [270] [271] Workforce losses, reduced tourism and trade, and increased health care and preparedness expenditures can have severe and long-term effects on economies.[272] [273] The World Bank (2014) estimated that the short-term economic impact of the 2013–2014 Ebola epidemic in Liberia, Sierra Leone, and Guinea would be between US $2.2 and $7.4 billion, with even greater long-term losses projected. These costs are compounded if a disease reaches a pandemic level, and the World Bank estimated a severe pandemic influenza could cost as much as $3 trillion globally.[274] [275] In the US, an H5N1 pandemic could cost the United States an estimated $165.5 billion in healthcare and medical costs, with much of the burden falling on private hospitals. While many industries suffer severe revenue and investment losses, some industries see increases in profits during disease outbreaks or epidemics due to an increased demand for medical products and services.[276] [277] Although economic losses from epidemics are widely studied, research on the converse financial effects of disease outbreaks is sparse.[278] This study investigates and explores potential beneficiaries from infectious disease events and the effect that infectious disease event spending may have on future infectious disease outbreaks.

Specialized Equipment and Infrastructure

During an epidemic or outbreak, the demand for personal protective equipment (PPE) grows rapidly in affected countries. Typically, companies that employ health care workers that are in close contact with infected individuals purchase PPE most often. However, during the 2003 Severe Acute Respiratory Syndrome (SARS) outbreak, a high level of public concern over sustained disease transmission created greater than expected demand for PPE among the general public. [279] [280] The overall increased demand of PPE during outbreaks, between government agencies, hospitals, and the public, may benefit the manufacturers and suppliers of personal protective equipment.[281] In 2014, PPE suppliers saw an initial uptick in demand for their products when the West African Ebola outbreak began receiving intensified media coverage. After news spread of the first Ebola infected patient in the US, Medline Industries, one of the largest PPE suppliers in the US, reported a 40 percent increase in domestic demand for facial-protection products and a 50 percent increase in demand for isolation gowns.[282] Medline increased production of these items and packaged them in specialized Ebola protection kits, which also included body suits, face shields, boot covers, masks, gloves and biohazard bags to meet the increased demand following the Dallas Ebola case. Much of the economic burden of these PPE purchases fell on US hospitals, many of which bought enough supplies to meet or exceed the revised Centers for Disease Control and Prevention (CDC) guidelines.[283] CDC's new 2014 guidelines on Ebola preparedness required selected hospitals to increase supplies of PPE and other safety equipment.[284] US hospitals spent money on Ebola preparedness PPE, including coveralls, gowns, face shields, and respirators. One hospital in Kansas reportedly spent $100,000 on jumpsuits, hoods, and respiratory devices. This is the estimated amount of PPE to treat one Ebola patient for two weeks. The State University of New York Upstate Hospital estimated short-term Ebola preparedness costs, including building modifications for containments, at $448,052, with long-term ongoing Ebola operational costs (e.g., biosuits, respirators, laboratory costs, waste disposal costs, operating costs, lost revenue, training, and planning) at an additional $50,000 per month. The North Shore Long Island Jewish Health System, a health care network of twenty-one hospitals, proposed building a biological containment unit to treat and contain infectious diseases within one of their facilities for an estimated cost of $15 million.

Additionally, when implementing new Ebola-preparedness guidelines, the CDC purchased $2.7 million worth of PPE, including gowns, coveralls, aprons, boot covers, gloves, face shields, hoods, N95 respirators, and disinfecting wipes, to create a strategic national stockpile. This equipment was divided into fifty kits, each able to treat one Ebola patient for up to five days.[285] Because there were so few Ebola cases in the US, critics questioned the necessity of CDC and hospital equipment stockpiling, which exhausted already depleted PPE manufacturer resources. DuPont, a producer of chemical suits, boot covers, face masks, and hoods, who works with the CDC and the Department of Homeland Security (DHS), struggled to fill demand in priority Ebola-stressed countries in Africa.[286] [287] Due to these surges in demand, companies like DuPont that were capable of producing and selling large quantities of PPE and other health equipment generally had stock spikes, increased sales, and designed new products specific to Ebola response. For example, Medline Industries reported over $7 billion in sales in 2014, up from $5.8 billion in 2013.[288] [289] Following a rapid increase in reported Ebola cases and international attention, BioMedical Devices received ten times as many orders in October and November than September of 2014, and their products went into backorder. Alpha Pro Tech (APT) and Lakeland Industries, both PPE suppliers, saw share increases of over 200 percent in

2014 (Figure 1).[290] Lakeland Industries attributed their gross profit increase of 9.3 percent, in the fourth quarter of the 2015 fiscal year (ending on January 31, 2015) to sales of PPE related to the US Ebola response. During this period, Lakeland tripled production capacity for PPE and obtained several large contracts.[291] Lakeland had to sell shares of common stock in October 2014 to support increased production due to the high demand of Ebola-related safety products, although the share price fell 26 percent after panic concerning the Ebola epidemic cooled slightly. Nonetheless, in fiscal year 2014 Lakeland's performance was higher than previous years, and the company ended the year with total revenue of $99,734,000 and a gross profit of $33,712,000. Similarly, Alpha Pro Tech stated that the 28 percent increase in their company's Infection Control business segment was primarily due to increased sales in the fourth quarter of 2014, a result of the US response to the Ebola epidemic.[292] The response of increased demand, production, and income in the PPE industry is not unique to the Ebola epidemic and is evident in other outbreaks like the SARS outbreak of 2003.

The 2003 SARS outbreak created similar demand surges for PPE and other protective products. Halyard Health of Kimberly Clark hit their PPE production cap during this SARS outbreak with orders rising 30 percent above normal levels.[293] Some companies responded to this increased demand for protective equipment during this outbreak by developing new and highly effective preventative and protective equipment. Singapore Technologies Electronics, a company of Singapore Technologies Engineering, developed a thermal camera system able to detect fevers and, within five weeks, sold 138 camera systems, contributing an estimated $12 million to the company's annual sales.[294] Similar to the Ebola epidemic, increased demand and sale of PPE during the SARS outbreak created unique opportunities for companies that provide necessary specialized equipment to profit from affected governments and treatment facilities.

Pharmaceutical Industry

The pharmaceutical response to pandemics is one of the most important factors in effectively preventing and eradicating the spread of infectious diseases.[295] While some research has supported a combination of pharmaceutical and non-pharmaceutical interventions as the most cost-effective mitigation strategy, others posit that increased vaccine and drug course availability alone may reduce economic losses most effectively.[296] An effective vaccine response during an outbreak can significantly limit the spread and impact of disease spread within the affected country, which ultimately may mitigate in-country economic losses.[297] Large amounts of federal funds are invested in the development and provisioning of vaccines to hospitals and the public during an outbreak. Increased demand for medicines and vaccines during an outbreak often leads to overall economic gains for the healthcare and pharmaceutical companies that produce them. Companies involved in influenza medicine or vaccine production often see greater profits during years with influenza outbreak or pandemic events. After the 2009 H1N1 influenza pandemic, Sanofi-Aventis, now known as Sanofi, reported that their net profits of 7.8 billion Euros were a record high due to the production and sale of influenza vaccines.[298] The value of the worldwide market for influenza preventative products and treatments alone has been estimated at $2.5 billion, with the flu antiviral market valued around $1.5 billion worldwide, and this market is mostly controlled by Roche, Sanofi-Pasteur, GlaxoSmithKline, and Novartis.[299]

For these large pharmaceutical companies, vaccine sales are rarely as profitable as drugs sales, which can lower the incentive to develop them.[300] Prior to the 2014 West Africa Ebola epidemic, Ebola vaccine development was not considered profitable by pharmaceutical or medical research companies because outbreaks were considered too rare and limited in scope.[301] This perception may be changing, as developing countries become new viable markets for vaccines.[302] The global market for vaccines in 2014 is estimated at $25.5 billion, and approximately $6.1 billion of $7.1 billion federal avian influenza funds in the United States were earmarked for vaccine development, production, and stockpiling.[303] Severe outbreaks like the 2014 West Africa Ebola epidemic have incentivized vaccine development and contributed to making vaccines one of the fastest-growing areas of research within the biotechnology industry. Collaboration between governments and pharmaceutical companies is necessary to develop affordable and accessible vaccines for the next outbreak or epidemic. Strategies like government stockpiling can help ensure supplies of vaccines are adequate in preparation for an infectious disease event while minimizing purchasing costs.

Drug Stockpiling

Stockpiling medicine can be an economically efficient response to a pandemic.[304] A cost benefit analysis revealed that stockpiling investments for influenza are economically cost effective if there are at least two influenza pandemics every eighty years. However, overestimation or mismanagement of federal stockpiles can lead to significant federal economic losses. In a 2014 audit of DHS, the department was criticized for their poor management of Influenza vaccine and PPE inventory and their lack of stockpile replenishment plans for expired vaccines and equipment. Stockpile replenishment plans are especially important, as 81 percent of DHS's Influenza antiviral vaccine stockpile will expire by the end of 2015, leaving the US vulnerable to an Influenza outbreak.[305] DHS has also been called into question for their decision to place a $463-million-dollar order with Siga Technologies for an expensive Smallpox vaccine to counter bioterrorism. It is suspected that the drug's $200 per treatment purchasing cost is a significant markup from its production cost, and the two-million-dose government order significantly increased the small company's profits.[306]

As was the case with Siga Technologies, drug stockpiling has significantly increased revenues for many pharmaceutical companies. After continuous outbreaks of H5N1 starting in 2003, the World Health Organization (WHO) amassed a stockpile of oseltamivir treatment courses that were ready for use by 2006.[307] In 2005, Hoffmann La-Roche, a private pharmaceutical company, and the main producer of oseltamivir in the form of Tamiflu, reported it was their best year ever partially due to these Tamiflu stockpile sales.[308] Relenza, a form of zanamivir sold by GlaxoSmithKline, was also internationally stockpiled in response to the H1N1 pandemic threat in 2009, and the company saw revenues of approximately $1.4 billion in pandemic vaccine sales alone.[309] [310] Millions of these treatment courses were purchased by the US government to be included in the CDC's Strategic National Stockpile that was amassed in response in the 2009 H1N1 pandemic and maintained in the event of future influenza outbreaks. In 2009, estimated total sales of influenza vaccines and adjuvant totaled US $6.9 billion, according to JP Morgan. The burden of these stockpiling costs fell mostly on governments that increased stockpiles as part of pandemic preparedness programs.

Hoffman La-Roche also provided treatment courses in Vietnam as the government increased vaccine stocks as part of the 2005 H5N1 and Human Influenza Pandemic Preparedness Plan. The Vietnamese government purchased 2.5 million treatment courses (twenty-five million capsules) of oseltamivir from Roche to be added to their stockpile.[311] The Thai government similarly purchased 260,000 treatment courses (2.6 million capsules) of oseltamivir from Roche for a national stockpile throughout 2005 and 2006, and enacted policies to increase the domestic production of oseltamivir each year by 100,000 treatment courses over the course of three years.

Manufacture Reserve Programs initiated by Roche and GlaxoSmithKline in 2008 charged hospitals and other private organizations in the program an annual fee to reserve the ability to buy treatment courses in the case of a future influenza outbreak. These programs have further inflated profits for drug manufacturers. While pharmaceutical companies see an increase in profits due to outbreaks or intensification of government pandemic preparedness programs, the success is often short lived, and vaccine markets are exceptionally volatile during disease outbreaks. This is reflected in fluctuations of stock market prices.

Stock Prices & Money Markets

Changes in stock prices during outbreaks illustrate market fluctuations within pharmaceutical and biomedical research sectors during periods of intensified pandemic preparedness or heightened perceived pandemic risk. The recent West Africa Ebola epidemic of 2014 led to a huge push in medicine and vaccine development. The first case of Ebola in the US caused stocks for many pharmaceutical companies to soar. One of the biggest market gainers was Tekmira Pharmaceuticals Corporation, now known as Arbutus Biopharma Corporation, who developed the Ebola drug TKM-Ebola under a $140 million contract with the US Department of Defense. TKM-Ebola was fast-tracked for use in West Africa in 2014 and shares in the company jumped from $4.11 CAD per share to $27.85 CAD (17.31 percent increase) on the Toronto Stock Exchange.[312] Despite these financial gains, TKM-Ebola was found ineffective, and the company suspended development and changed its corporate name to Arbutus Biopharma.[313] Once Sarepta Therapeutics (SRPT) began developing an Ebola treatment with a high trial success rate, company shares went up from $1.67 to $22.77 in after-hours trading, eventually closing at $21.10. Other companies involved in Ebola vaccine or medicine development, including Inovio Pharmaceuticals and AstraZeneca, also had stock spikes in 2014, correlated to the timing of various events within the Ebola epidemic. As observed with Sarepta's stocks, spikes in pharmaceutical markets during outbreaks are often quick and drastic; however, these gains are often short lived (Figures 1-2).[314] Biotechnology companies during the SARS outbreak of 2003 also saw sudden, brief spikes in stock prices related to statements or evidence that their products could be useful against SARS. Shares of SciClone Pharmaceuticals rose from $5.56 to $6.30 within a short period in May 2003, and the company saw a definitive increase in sales during this time (Figure 2). Stock returns of Taiwan's biotechnology sector also had positive surges in stock returns in response to the SARS out-break. While disease outbreaks may be beneficial to biotechnology markets, other markets, like those for non-essential goods, can suffer severe losses as a result of panic or perceived risk among consumers. Outbreaks also affect financial trading behaviors. Professional finance companies use their exten-sive knowledge of these trends and industry expertise to profit during dis-ease outbreaks. In 2014, hedge funds predicted that Ebola would affect the Ivory Coast, a major cocoa producer, and profited from their bet that cocoa prices would continue to rise.[315] However, this price volatility did not nec-essarily reflect an actual shortage in the cocoa supply chain and may have been a symptom of market speculation.[316] Similarly, a 2005 Citigroup report warned investors about investing in labor-intensive industries and countries with inflexible labor laws during avian influenza outbreaks because, in the event of decreased demand, laborers cannot be laid off.[317]

Several companies like Natixis Global Asset Management and BMO Nesbitt Burns have produced investor guides for avian influenza and other diseases with pandemic potential.[318] [319] Visiongain, a business intelligence provider, advertises itself as being able to identify, examine, and provide timely consultancy on sectors with profit potential, including the pharma-ceutical and vaccine markets during pandemics.[320]

Trade

Increasing globalization and trade liberalization over the past twenty years has greatly expanded the world economy, allowing new opportunities for both developed and developing countries. For most countries, international trade as a percentage of country gross domestic product (GDP) has consis-tently increased each year.[321] Disease outbreaks and epidemics cause market volatility, particularly in the most economically integrated, globalized coun-tries where trade is a higher share of total GDP. As dependence on interna-tional trade continues to increase, countries are left vulnerable to the market fluctuations during epidemics. These fluctuations stem from many factors, including consumer fears of health risks of traded goods, and international trade regulations limiting dissemination of goods from outbreak countries. In South Korea, overseas shipments fell 10.9 percent in 2015 compared to the previous year, most likely in reaction to the outbreak of MERS in Seoul. Outbreaks of zoonotic diseases can cause severe market shocks, and trade income is lost because of decreased demand or trade bans on livestock and poultry goods. This was evident in the Bovine Spongiform Encephalopathy

(BSE) and new variant Creutzfeldt-Jakob disease (nvCJD) outbreaks in England, which caused \$5.75 billion in total losses, with \$2 billion lost in beef exports alone, and the outbreak of Nipah in Malaysia, which lost an estimated \$120 million in pork exports.[322] [323]

Economic losses from outbreaks extend beyond the outbreak country markets and affect the economies of trade partners as well. These secondary effects may be exacerbated in the future, as individual country economies become more dependent on foreign goods. China is the largest source of US imports overall, accounting for over 21 percent of total US imports in 2015.[324] China is a hotspot for disease emergence, and future disruptions to China's economy like the 1997 avian influenza outbreak or 2003 SARS outbreak would have significant effects on the US markets and consumption of these goods.[325] [326] Often markets fall as consumer confidence in a product declines. Sometimes, this can lead to increase in demand for an alternative good. After the 2004 outbreak of H5N1 avian influenza, prices of non-poultry meats rose 30 percent in Cambodia and Vietnam as demand increased for meat goods that were considered safer. This can translate into higher export income for trading partners that sell products to help meet the raised demand for alternative goods.

Furthermore, when demand remains but trade bans or production losses due to an outbreak limit a country's ability to meet the demand, competing markets may have the opportunity to increase their trade flow to replace the lost supply. Following the first case of BSE in the US, beef and cattle exports dropped and consumers of US livestock products, who instituted bans against the import of US goods, began looking to Australia, New Zealand, and South America to cover the shortage, allowing these markets to flourish.[327] A similar positive effect could have been seen in US poultry markets if outbreaks of Avian Influenza continue in Asian countries. If the US continued to remain free of Avian Influenza outbreaks, it was predicted that the US poultry industry may have expanded to cover the losses of the Asian markets. Therefore, disease outbreaks, or heightened perception of disease risk among consumers, can have both positive and negative effects on trade profits and sector revenues depending on the relation of the goods being sold to the specific outbreak (e.g., beef trade decreases during BSE outbreak, but can increase during Avian Influenza outbreak).

Other Benefiting Sectors

Changes in consumer sentiment during outbreaks have smaller scale effects too and can result in unusual success for specific products or services used by the public. The demand surges for PPE during the 2003 SARS outbreak in China were unusually galvanized by public consumers, and supermarkets and hypermarkets saw a short-term doubling of business through increased sale of disinfectant and hygiene products. Some businesses, like CK Life Sciences International, who marketed their health beverage drink Vitagen as helping fight SARS, saw new avenues for profit in this new market for preventative products. The Federal Trade Commission (FTC), the US Food and Drug Administration (FDA), and Margaret Chan, then Hong Kong's director of health, even criticized some companies for exploiting the high demand and consumer ignorance surrounding the SARS outbreak in China. The FTC released an official warning to online companies that marketed these fraudulent products and ordered removal of any claims or suggestions that their products would protect against, treat, or cure SARS without any scientific proof to support these claims.[328] Similarly, the insurance industry experiences gain directly related to outbreak fear. A representative of American International Group (AIG), an insurance company with a large presence in China, stated that the SARS epidemic would boost sales of life insurance and, in response, AIG would create products specifically catering to SARS concerns. Additionally, new types of insurance are being introduced for private businesses to mitigate pandemic-specific costs. In 2014, Miller Insurance Services and William Gallagher Associates (WGA) announced "Pandemic Disease Business Interruption Insurance," a coverage plan that responds to the direct loss of income due to quarantine events.[329] This insurance is a direct response to health care facility shutdowns and low revenues in the aftermath of the Dallas, Texas Ebola quarantine. Insurance companies providing pandemic insurance like Berkshire Hathaway, Catlin Insurance Company, Lexington Insurance Company and their Global Supply Secure Program, Munich Re, and Montpelier Re typically require an official government declaration, often a WHO pandemic alert level three to six.

There is also a growing market for international pandemic insurance coverage to facilitate the quick deployment of health workers and supplies for containment before an outbreak reaches pandemic-level threat. In response to the lag time in resource mobilization in 2014 for the West Africa Ebola epidemic, the World Bank began consulting with the African Union, the United Nations, and national governments to develop a Pandemic

Emergency Financing Facility (PEF) with an insurance scheme for affected countries and international organizations. The PEF will cover the costs of deployment of workers, medical equipment and supplies, pharmaceuticals, food, and coordination of efforts but not pandemic reconstruction costs. PEF funds will be available to countries and organizations like WHO, World Food Program, UNICEF, and Médecins Sans Frontieres. PEF's Private Insurance Mechanism will insure developing countries by buying insurance coverage from the private sector on their behalf. Projected risks to investors will likely make premiums for the Pandemic Emergency Facility expensive and therefore only useful when a large payout is needed.[330] Country-level pandemic insurance may be an economically efficient mechanism, as countries will have access to finances to control an outbreak or epidemic before there is pandemic threat and large sums of foreign aid is needed.

CONCLUSION

While the negative economic impacts of an epidemic are well documented, the positive economic effects of epidemics and outbreaks have been less well studied. This study found that in many cases organizations are profiting from epidemics and pandemics. Infectious disease outbreaks, epidemics, and pandemics place governments in a difficult position where already limited government funds must be reallocated to deal with imminent infectious disease threats. Interestingly, the redirection of funds to combat imminent infectious disease threats like Ebola and SARS is increasing society's preparedness for dealing with future infectious disease threats (e.g., stockpiling and research and development).

Future research should investigate the price of specialized healthcare equipment and supplies to determine when corporate profiteering conflicts with the public good (i.e., the pharmaceutical industry working with research scientists to develop medical countermeasures for infectious diseases created in a laboratory which do not exist in nature). This could lead to productive actions to guide policies and regulations for goods critical for public health, especially in emergency situations. Additional research should examine the cost-effectiveness of practices involving vaccine and anti-viral prophylaxis stockpiling to encourage the government to work more closely with the pharmaceutical industry to develop agile and flexible production systems to maintain corporate profitability while reducing wasteful spending by the government. Lastly, the government should use a systems-based approach to build capacity for preparedness. For example, PPE producers had to meet increased production demand and were running low on necessary production resources during the 2014 West Africa Ebola epidemic. By working more collaboratively and closely with the PPE or pharmaceutical companies that produce lifesaving drugs there is extensive opportunity to create effective policies that increase public health preparedness for the next pandemic threat and industry plays a critical role in this process. Corporations making profits is not a problem, but the knee jerk reaction to emerging infectious disease threats ends up costing society more than it must.

CHAPTER 22

Where Do We Go from Here?

There are many things we can do to prevent a man-made pathogen from causing another pandemic.

The first is a complete international ban of gain of function research except in the limited cases where it is necessary to build live attenuated vaccines, which have been highly effective in preventing diseases.

Second, we need to eliminate the concept of Dual Use Research of Concern (DURC) from all US policy on infectious disease research. Any biological research that designs or genetically engineers a pathogen via serial passage or genetic engineering to enhance its transmissibility, pathogenicity, virulence, infectivity, or survivability is incredibly risky and dangerous.

With today's synthetic biotechnology, the concept of select agents is outdated and should not be a factor in evaluating the risks of gain of function work.

Just imagine a scientist taking some of the deadliest foodborne pathogens and genetically engineering them so that they were transmitted via aerosols and the havoc that could cause to society.

Yet, this type of GoF work would not be considered high risk because the foodborne agent being modified is not on the select agent list.

For these reasons, the stated intent of the researchers does not matter, nor does it mitigate the risks associated with GoF work. If anything, the stated intent of GoF scientists and engineers is a representation of their biases toward GoF being safe, and the false sense of security pertaining to the highly dangerous work that they are engaged in.

Therefore, the concept of DURC needs to be eliminated and we need policies that specifically focus on GoF. The concept of DURC can apply to anything, and that is the problem with using the term in policy.

The current DURC policy is unenforceable, and the people that created and enforced the policy were heavily biased to DURC being safe, which ultimately led to a handful of corrupt and greedy scientists killing millions of people globally.

These facts are undeniable.

If GoF is to exist at all (beyond the scope of live attenuated vaccines), then GoF should only exist domestically in highly secure and tightly controlled laboratories within the national laboratory system or the Department of Defense. The technocrats leading government organizations, like Dr. Anthony Fauci, need term limits to prevent them from massing power which enables them to circumvent Congress.

Though he is finally stepping down at the end of 2022, almost forty years of one man as director of NIAID is too long. It is unbelievable to me that we allowed academics and bureaucrats to self-determine and self-regulate what constitutes a biological weapon.

We can no longer trust or allow people like Dr. Peter Daszak, Dr. Ralph Baric, Dr. Anthony Fauci, or other academics to self-regulate.

I hope that you enjoyed reading this book, and it motivates you to act to stop this madness. I want to leave you with my favorite rejected publication. The following article provides a public health theory to effectively eliminate the burden of global disease.

We can't improve public health without first hypothesizing, measuring, and comparing what things improve public health the most. The pharmaceutical industry wants you to believe that the most impactful and effective way to improve health is via pharmaceutical intervention.

While medical innovations and technology are certainly important to public health, they are far less important than the pillars of public health. We need to properly focus our public health efforts and here is a theory, a set of ideas, of where we should focus our efforts.

A Theory of Public Health Necessity

Abstract
Since the World Health Organization (WHO) was founded in 1948, there has been little reduction in the rate of infectious disease emergence despite success stories in controlling specific diseases. This manuscript is motivated

by a need for effective and comprehensive approaches to implementing mitigation strategies and policies. The hierarchy of basic public health needs described in this text is founded in experiments and empirical evidence derived from numerous public health studies that have examined the effects of various exposures (e.g., energy, vaccines, clean water, shelter, etc.) and their associated public health outcomes (e.g., morbidity, mortality, disability adjusted life-years (DALY), quality adjusted life year (QALY), potential years of life lost (PYLL), global burden of disease (GBD). For a global public health system to function most effectively, basic health needs must be fulfilled (i.e., potable water, food and nutrition, shelter, energy, sanitation, education, health care, medical technology). We view public health components as a hierarchy of needs and associated resources required to maintain human health starting from the most basic needs that would secure the health of billions of people and progressing vertically toward more complex systems and interventions to benefit thousands or millions. Each hierarchy level builds on previously established factors and thus incrementally improves and prolongs quality and duration of life.

Introduction
Public health strategy, policy, and investment are not adequately preventing global disease. There is insufficient access to potable water, nutritious food, shelter, energy, sanitation, and education globally. These needs have the most impact on global health, yet most global public health strategy, policy, and investment focus on healthcare and medical technology and have had little impact on reducing the overall global burden of disease. The World Health Organization's International Health Regulations (IHRs) were a positive step toward improving global health by improving the ability to detect and respond to diseases, but are deficient in preventing infectious disease outbreaks from occurring in the first place.[331] The IHRs focus on what this article refers to as post-outbreak healthcare and medical technology, rather than more fundamental and impactful areas of public health, which align more closely with the Sustainable Development Goals. If all countries globally were providing the underlying requirements for healthy societies, then the IHRs would be the logical next step for improvement; however, each country and community has different needs requiring attention to prevent disease. Given the variation in the levels of core public health requirements, a logical framework can help to determine which public health needs should be addressed first. This framework should focus on preventing disease from occurring, not responding to it. This manuscript proposes a general theory

of public health needs like the "Theory of Human Motivation" proposed by Maslow.[332] This manuscript is an attempt to formulate a theory of public health necessity that satisfies theoretical demands while conforming to global public health's known facts, and experimental, clinical, and observational health research. With this logical framework, global public health strategy, policy, and investment should be guided. Exporting public health solutions for higher order problems to areas where more basic challenges predominate are often unsustainable at best. The purpose of this framework is to maximize the impact of resources to best improve global public health by viewing eight key areas as essential building blocks: potable water, food and nutrition, shelter, energy, sanitation, education, health care, and medical technology (Figure 1).

Potable Water

Access to potable water is the most important public health need and the cornerstone of improved sanitation and biosecurity. The combination of clean water, adequate sanitation, and sufficient hygiene has the potential to prevent at least 9.1 percent of global disease burden and 6.3 percent of all deaths worldwide.[333] WHO defines safely managed drinking water in a household as being located on the premises, available when needed, and complying with fecal and chemical standards.[334] While access to potable water is considered a basic human right, in 2015 one in ten people still lacked access to safe water globally.[335] Increased access to clean water is necessary to prevent diseases contracted from contaminated water, including fecal-oral route transmission diseases like diarrhea, and is required in the treatment of many diseases. Additionally, clean water for sanitary purposes is integral to preventing disease transmission because personal hygiene (predominantly handwashing) is the only absolute protective barrier for blocking fecal-oral routes of disease transmission.[336] Studies show that washing hands with soap can decrease the incidence of diarrhea, severe intestinal infections, shigellosis, and pneumonia in children by approximately 50 percent.[337][338] Universally improved drinking and washing water has the potential to reduce contraction of diarrhea by 45 percent and morbidity by 21 percent worldwide,[339] indicating the gap in providing this basic public health need. Diarrhea alone is a significant health issue: diarrheal disease remains a leading cause of child morbidity and the second greatest cause of mortality in children under five years old in the world.[340] Access to fresh drinking water is limited by the availability, management, and equitable distribution of freshwater sources, as well as sanitation infrastructure. Universal access to potable water has beneficial secondary effects to public health. For example, diarrhea and other waterborne illnesses are a principal reason for both school and workplace absenteeism.[341] Increasing access to potable water therefore produces concomitant benefits for education and worker absenteeism. Additionally, improved water facilities enable more children, particularly girls, to attend school because of the reduced burden for time spent collecting water.[342] [343] [344] These secondary effects further reduce disease burden and will be discussed in subsequent sections in order of their relative importance (Figure 1).

Food and Nutrition

Access to sufficient and nutritious food is imperative to human survival and disease prevention, yet a significant proportion of the human population is not food secure.[345] [346] [347] [348] Food insecurity is the inability to acquire nutritious food through socially acceptable means and significantly contributes to poor health.[349] Adults in food insufficient households are more likely to have unhealthy, nutrient-deficient diets, and food insecurity is associated with an increased rate of chronic illnesses including diabetes, heart disease, and high blood pressure.[350] When food insecurity develops into malnutrition, health risks increase.[351] Malnutrition and extreme food deficiencies significantly impair the health and development of individuals. The cyclical relationship between malnutrition and immune response dysfunction makes malnourished individuals particularly susceptible to infection and disease. Even mild forms of malnutrition are correlated with stunted growth and higher mortality rates.[352] [353] Approximately one third of children in developing countries under the age of five years are malnourished or suffer from chronic malnutrition, and are subsequently burdened with health issues like stunted growth, physical wasting, and being underweight. The Dutch Famine of 1944–1945 highlights the lifelong impact of malnutrition on human health: the malnutrition of pregnant women during the famine led to a higher rate of severe chronic illnesses in their offspring several decades later.[354] In the developed world, malnutrition can occur in individuals who do not eat proper quantities of nutritious food. Residents of low-income neighborhoods often lack access to affordable, healthy food suppliers, and their resulting diets are unhealthy and nutrient-poor, leading to a high prevalence of obesity, diabetes, heart disease, and other chronic conditions.

Micronutrient malnutrition is responsible for a wide range of non-specific physiological impairments, leading to reduced resistance to infections, metabolic disorders, and impaired physical and psychomotor development.[355] [356] Specific nutrient deficiencies are also correlated with the increased risk of certain diseases, often through a reduction of the mucosal immune function. Cost efficient interventions could mitigate these health risks. For example, research suggests that vitamin A-fortified golden rice could halve the impact of vitamin A deficiency on the population of India while maintaining cost effectiveness.[357] A barrier to growing nutritious food is access to energy and electrification. Crop yields in low-income countries are currently much lower than those of farmers in high-income countries, who can use modern agricultural techniques.[358] Research on energy-sustainable agriculture that does not require intensive, industrial-level resources could alleviate food insecurity in developing countries.

Shelter

Access to adequate and healthy shelter is a basic health need. Despite this, nearly 40 percent of urban growth has occurred in slum housing.[359] Inadequate housing is defined as housing with moderate or severe structural problems including deficient plumbing or electricity.[360] Adequate housing protects against communicable and chronic diseases, injuries, poisonings, and both psychological and social stress.[361] [362] [363] [364] Unhealthy housing is defined as exposure to toxins or detrimental environmental conditions within the house, and includes exposure risk to rodents, mold, and water leaks. Both inadequate and unhealthy housing can leave residents vulnerable to chronic and communicable health threats.

Inadequate shelter conditions often exist in a context of poverty, warfare, and limited infrastructure.[365] [366] [367] Studies on refugee camps and homelessness highlight the health effects of poor housing conditions. Refugee camps are often overcrowded and plagued by inadequate heating, dampness, molds, poor lighting, and poor ventilation and are associated with incidence of respiratory diseases and fever.[368] [369] Bonner et al. found that households in Liberian refugee camps in Sierra Leone with more rat burrows and poorer external hygiene had a higher incidence of Lassa fever cases compared to nearby houses that were in better condition. Furthermore, housing instability can lead to limited access to medications and health care, and increased rates of chronic diseases, hospitalizations, and emergency department use. Clean household energy is a primary component of healthy housing. Nearly half of the world's population lacks access to clean household energy and must burn unprocessed biomass fuels like wood, coal, dung, and crop residues to heat the household and cook, which can lead to unhealthy housing scenarios.[370] [371] When compounded with poor indoor ventilation, the use of unprocessed biomass fuels indoors creates indoor air pollution (IAP) that causes 2.7 percent of the global disease burden. Nearly half of the world's population has IAP, with women and children at highest risk of exposure. IAP-associated conditions include acute lower respiratory infections, chronic obstructive pulmonary disease, lung cancer, cardiovascular disease, and other health conditions.[372][373]

Although the exact size and nature of health gains from housing improvements are unknown, studies suggest that provisioning housing to individuals living without adequate shelter improves their self-reported physical and mental health, decreases rates of substance abuse, and increases health service use.[374][375] Because of the association between poor housing conditions and disease risk, providing stable housing may alleviate associated health issues. Providing adequate shelter should be a fundamental concern of building successful public health infrastructure on which other key policies can be built. Policy intervention must work at multiple levels to solve this fundamental public health need by addressing intersecting factors including the provisioning of clean energy.

Energy

Accessible, clean energy has the potential to improve and modernize public health systems at multiple levels. Studies have revealed direct positive correlations between energy consumption per person, life expectancy, and infant mortality.[376][377] Clean energy can power water and sanitation infrastructure and mitigate harmful exposure to indoor air pollution from lighting and cooking with biofuels.[378] Unfortunately, access to clean energy is not equitably distributed. Between two and four billion people depend on biomass fuels and are vulnerable to their direct health consequences, while lacking access to the indirect benefits of clean energy (e.g., improved transportation, agriculture, and health care). Policy should stimulate communities to move up the energy ladder by transitioning from burning biofuels to modern electric grids that would reduce both energy costs and pollution. Energy reforms that help provide clean and accessible energy to more of the global population can significantly improve the health conditions of households and health care facilities. A major benefit of clean energy and electrification is reducing exposure to indoor air pollution created by burning biofuels. Additionally, all levels of the health care industry, from community clinics to regional hospitals, benefit from improved electric infrastructure. Access to electricity can facilitate the use of better laboratory and diagnostic technologies, provide light for medical services during nighttime, ensure better sterilization of equipment, and allow the management of thermosensitive treatments (e.g., vaccines).[379] [380]

Electrification policies must be constructed within their cultural context to ensure their wide adaptation. Globally, many families prefer to cook on biofuels even when electricity is available, leaving low-hanging unrealized health benefits on the table. Two common methods to increase electrification are extending grid access to communities and setting up decentralized energy production facilities, primarily based on renewable energy. For example, in Liberia more primary care providers use solar power than fossil fuels, which has concomitant benefits for other basic global health necessities like clean air and potable water. Additionally, the electrification of health care facilities and households allows for access to information communication technologies that can radically improve health information and data systems and disseminate important health care information through TV and radio.[381] Access to information communication technologies can revolutionize education for both health care workers and the general public. In these ways, clean energy and electrification are building blocks for the development of health care technologies, improved access to health care, better educational tools, and sanitation system infrastructure.

Sanitation
Sanitation measures that adequately separate human and animal waste from human contact are a primary tool in achieving safe and clean water supplies.[382] Proper management of solid waste, management of clean water, and promotion of hygiene practices are effective means to limit the spread of disease. The primary effects of poor sanitation, hygiene, and contaminated water are water-borne diseases like cholera, viral hepatitis, typhoid, and other diarrheal diseases. In total, 4.3 percent of the global disease burden is attributed to diarrheal diseases, with 88 percent estimated to be caused by unsafe water and inadequate sanitation and hygiene.[383] Effective management of waste should be a top public health priority on which other systems can be built.

Urbanization has posed challenges to sanitation management throughout history, which continues into the modern developing world.[384] The increased spread of infectious diseases like cholera during the urbanization and industrialization of nineteenth century London illustrates these challenges. Britain was a pioneer in passing legislation and setting up governing bodies to reform sanitary conditions in the overcrowded, newly industrial cities in response to outbreaks like the 1847 cholera epidemic.[385] [386] Edwin

Chadwick's research definitively linking environmental conditions to disease incited the Public Health Acts of 1848 and 1875 that outlined plans for improved drainage and sewer systems, and regulated waste disposal practices.[387] These acts were important precedents to government involvement in urban waste management.

Despite the increasing role of government in waste management in many parts of the world, proper management of solid waste continues to be a global problem.[388] [389] [390] The unprecedented 828 million people living in urban slums in the developing world are without access to proper sanitation, which poses a grave human health risk.[391] In 2015, one third of the world's population did not have access to improved sanitation, with one eighth of the population practicing open defecation.[392] [393] Improperly disposed solid waste is often thrown into open spaces where stagnant water from clogged drains, refuse, and leachate percolating into waterways drives potable water contamination and disease outbreaks.[394] [395] [396] [397] [398] Additionally, home incineration and disposal of hazardous and biomedical wastes cause health problems by polluting groundwater and air.[399] [400] As rapid urbanization continues, sanitation and solid waste disposal will become a larger public health threat that calls for political solutions.[401] Successful sanitation policy combines technological implementation with increased cultural awareness.[402] The developing world's priority is the physical separation of waste from human populations, and a commensurate 90 percent of their municipal solid waste budgets are spent on this task alone.[403] [404] The developed world, having to a large degree accomplished this goal, may focus on improving waste processing methods and thus use less than 10 percent of their budgets for solid waste collection. To avoid deleterious public health outcomes, the developing world must start allocating funds to processing solid waste and transitioning from open landfills without treatment to sanitary landfills.[405]

Additionally, proper waste management and sanitation can have positive secondary effects on other public health goals. Strong and effective waste management policy directly aids local employment, energy production, the environment, and food production systems.[406] Sanitation reforms benefit women's health by attenuating physical violence in shared latrines and increasing access to education in schools that can accommodate menstruating students. Sanitation reform can also effectively regulate the sustainable production of nutritional foods. Wastewater is often used in crop irrigation and can cause increased risk of viral, bacterial, and protozoan enteric infections if improperly managed.[407] [408] [409] One policy solution is to provide strict guidelines on the maximum concentration of excreted pathogens (e.g., viruses, bacteria, helminth eggs, and fecal coliforms) in the wastewater used in agriculture to prevent transmission of communicable disease through this pathway. Because water and waste guidelines are inter-connected foundations to population health, a systems approach should be taken in the formation of waste and water policy.

Education

Public health education, including education on sexually transmitted dis-eases, smoking tobacco, and unsafe drug usage, is integral in disease preven-tion and outbreak responses to global health threats.[410] [411] Limited access to educational tools and services can impede knowledge and understanding of behavioral health hazards. For example, in 1996, two-thirds of adult Chinese smokers believed cigarettes did "little to no harm". Education on nutrition, sanitation practices, vaccines, antimicrobial resistance, and communicable disease transmission gives individuals the opportunity to make informed decisions about both personal and community-level disease prevention.[412]

The foundation for education is general literacy. Rajan, Kennedy & King found that increased literacy is one of two major factors (the other being alleviation of poverty) needed to improve public health throughout disparate populations. Literate adults can understand official written public health information like medical pamphlets, outbreak signage, or warning labels on tobacco products. Even if the adult population is mostly illiterate, educating children indirectly teaches parents, so education policy may be best suited focusing on younger generations.[413] This underscores the impor-tance of other foundational health needs, like electricity, clean water, and sanitation that contribute to decreased absenteeism from school and expand the opportunity for females to receive an education.

Health Care Systems

Once the public's health needs have been met, and education and literacy systems are in place, policies should focus on training and building the human and technological capital necessary to increase health care access.[414] Building a trained health care workforce is a first step in the development of a health care system. Necessary human capital in the health care sector is built through appropriate education programs and long-term retention of qualified health care professionals.[415] [416] [417] [418] The absence of strong accred-itation and training programs are an obstacle policy-makers must address to ensure a baseline level of health care competency worldwide.[419] [420] [421] [422] Skilled worker retention can be a significant impediment to building a health care system, particularly in rural areas. Improving basic infrastructure (e.g., access to shelter, potable water electricity, and sanitation) improves the quality of life for health care professionals and encourages their retention.[423]

Supplying trained workforces with necessary tools and medicines to treat patients can be challenging when reliable medical supply chain infrastructures are not in place. Developing countries frequently experience local and national drug shortages in which patients cannot access lifesaving drugs.[424] In some countries, routine, and cheap equipment, like chest tubes for trauma victims or ultrasound gel, are not sufficiently stocked in medical supply reserves and frequently run out.[425] Policy should aim at emboldening the private and public sectors to collaborate to create reliable supply chains for critical drugs and medical supplies.

Medical Technology
Government organizations can contribute to improving the overall health care system of their country through research, education, financing, and technology development. An educated and trained workforce within a sustainable healthcare system requires sustained funding to develop and thrive. Scientifically informed allocation of funds to treatment and vaccine research programs and the development of medical devices and biosurveillance systems is necessary to advance public health systems.[426] [427]

Lastly, government health care financing is a political opportunity to increase access and quality of health care for those least able to afford it. Exorbitant health care costs have forced 100 million people worldwide below the poverty line, while further impoverishing 1.2 billion of the world's poorest.[428] Furthermore, limited geographic access to health care facilities (e.g., primary care facilities and emergency medical services) can be an additional access obstacle, particularly in dense or rural isolated areas where there are transportation limitations and road infrastructure deficiencies.[429] [430]

Conclusion
In recent decades, public health policy and investment has focused on medical technology and healthcare and has not reduced global infectious disease risk. Instead of focusing on health care and medical technology, global public health policy and investment should focus on adequate accessibility to (from most to least important): potable water, safe food, safe and secure shelter, clean and reliable energy, sanitation, and education. Public health encompasses a broad variety of scientific and political fields in which many players and moving parts must collaborate to achieve optimal health.

Actions implemented in one area of public health necessity can have positive or negative effects on the capabilities of other sectors (Figure 1). Many factors outside the proposed framework may have impacts on the development of effective public health measures, and these needs will likely vary based on location. At a minimum, a highly effective and functioning public health system needs these simple proposed factors to reduce a multitude of negative health outcomes. Culture, politics, geography, political instability, human rights, and many other factors outside the direct influence of the traditional health sphere and the theory proposed in this manuscript can make or break a public health system. An example of this is women's rights, a factor that can have an enormous effect on the health of a population's women and children.[431] Severe injury and disfigurement, high rates of infection (in particular HIV transmission), and female/mother-infant mortality are not uncommon in cases of rape, female genital mutilation, acid-throwing, and other forms of abuse.[432] [433] [434] Political representation, economic equality, and education are all necessary to improve the health of women around the globe.

Country stability and peace are also important in overall public health. Areas with high levels of corruption and conflict suffer from the destruction of shelters, displacement of large groups of people, and deteriorating health-related infrastructure, including unstable water and food sources.[435] Research indicates that countries with worse scores on the Corruption Perceptions Index (CPI) had higher rates of maternal mortality due to a lack of equitably accessible health services or transparent public health organizations.[436] Chaotic, conflict-induced conditions leave populations susceptible to disease outbreaks and unable to access health care when inflicted.[437] Although these factors are not immediately vital to human health, public health organizations should consider the many overarching health implications involved in these issues and how they relate to the foundational needs outlined in this hierarchy.

My proposed approach to sustainable public health is consistent with and underscores the importance of the Sustainable Development Goals. While the burden of effort required for implementation of the IHRs falls on the public health community, far more positive health outcomes will result from the multi-sectoral shared commitment to achieving the Sustainable Development Goals. As illustrated in my hierarchical model, public health is far more dependent on sound public services than it is on medical innovation.

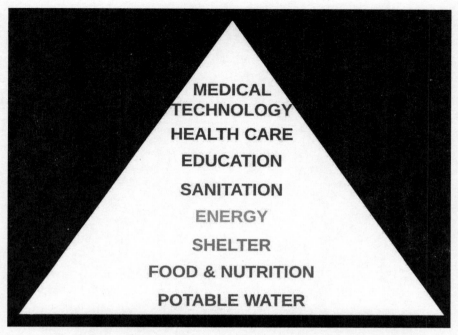

Figure 1. Contributing factors to a successful public health infrastructure. Factors are ranked from most fundamental and important to public health (starting at the bottom), with subsequent levels of increasing sophistication and decreasing importance to public health upwards.

Endnotes

Chapter 1

1   Alex Berenson, "Peter Daszak, Man of Many Hats," Unreported Truths (Substack), October 6, 2021, https://alexberenson.substack.com/p/peter-daszak-man-of-many-hats/comments.

2   Jeff Carlson and Hans Mahncke, "Covid-19 Outbreak & Cover-up," *The Epoch Times*, https://www.theepochtimes.com/infographic-covid-19-outbreak-acover-up_4025512.html?utm_source=ai&utm_medium=search.

3   Jo Ohm, "Leaky vaccines promote the transmission of more virulent virus," *Epidemics*, January 14, 2016, http://epidemics.psu.edu/articles/view/leaky-vaccines-promote-the-transmission-of-more-virulent-virus.

4   "Bayer Executive Says mRNA Vaccines are Gene Therapy," *We the Pundit*, March 8, 2022, https://wethepundit.com/bayer-executive-says-mrna-vaccines-are-gene-therapy-principia-scientific-intl-3/.

5   "Bayer Pharmaceuticals Division President Stefan Oelrich Brags That mRNA Shots are 'gene therapy' Marketed as 'vaccines,'" *Adverse Reaction Report*, March 4, 2022, https://adversereactionreport.com/news/bayer-pharmaceuticals-division-president-stefan-oelrich-brags-that-mrna-shots-are-gene-therapy-marketed-as-vaccines/.

6   Julia Mueller, "Fauci on COVID conspiracy theories: 'What we're dealing with now is just a distortion of reality,'" *The Hill*, August 23, 2022, https://thehill.com/policy/healthcare/3611899-fauci-on-covid-conspiracy-theories-what-were-dealing-with-now-is-just-a-distortion-of-reality/.

7   Kevin Breuninger, "Fauci blasts 'preposterous' Covid conspiracies, accuses his critics of 'attacks on science,'" CNBC, June 9, 2021, https://www.cnbc.com/2021/06/09/fauci-blasts-preposterous-covid-conspiracies-accuses-critics-of-attacks-on-science.html.

8   Evan J. Anderson, et al., "Safety and Immunogenicity of SARS-COV-2 mRNA-1273 Vaccine in Older Adults," *New England Journal of Medicine*, September 29, 2020, https://www.nejm.org/doi/full/10.1056/NEJMoa2028436.

9   "Safety and Immunogenicity Study of 2019-nCoV Vaccine (mRNA-1273) for Prophylaxis of SARS-CoV-2 Infection (COVID-19)," *ClinicalTrials.gov*, February 25, 2020, https://clinicaltrials.gov/ct2/show/NCT04283461.

10 Peter Doshi, "Covid-19 vaccine trials cannot tell us if they will save lives," *The BMJ*, October 21, 2020, https://www.bmj.com/company/newsroom/covid-19-vaccine-trials-cannot-tell-us-if-they-will-save-lives/.

11 "Moderna's COVID-19 Vaccine Candidate Meets its Primary Efficacy Endpoint in the First Interim Analysis of the Phase 3 COVE Study," *Business Wire*, November 16, 2020, https://s29.q4cdn.com/435878511/files/doc_news/2020/11/16/modernas-covid-19-vaccine-candidate-meets-its-primary-efficacy.pdf.

12 True North (@TrueNorthCentre), "Watch as Pfizer executive Janine Small admits to EU parliament," Twitter, October 11, 2022, 9:43 a.m., https://twitter.com/TrueNorthCentre/status/1579830040858329089.

13 Susan Jones, "CDC's Definition of 'Vaccine' Has Changed Over Time: 'Protection' vs. 'Immunity'," *CNSNews*, January 25, 2022, https://www.cnsnews.com/article/national/susan-jones/cdcs-definition-vaccine-has-changed-over-time-protection-vs-immunity.

14 "Immunization: The Basics," *Centers for Disease Control and Prevention*, September 1, 2021, https://www.cdc.gov/vaccines/vac-gen/imz-basics.htm.

15 Aled M. Edwards, et al. "Stopping pandemics before they start: lessons learned from SARS-CoV-2." *Science* 375.6585 (2022): 1133-1139.

16 https://promedmail.org.

17 "List of intelligence gathering disciplines," Wikipedia, last modified April 6, 2022, https://en.wikipedia.org/wiki/List_of_intelligence_gathering_disciplines.

18 Griselda González-Cardoso et al., "PM2.5 emissions from urban crematoriums," *Research Gate*, October 2018, https://www.researchgate.net/publication/328891567_PM25_emissions_from_urban_crematoriums.

19 Paul Bolstad, *GIS Fundamentals*, 6th Edition, XanEdu, https://paulbolstad.net/gisbook.html.

20 Redlands, C. E. S. R. I. (2011). ArcGIS Desktop: Release 10.

21 Ken Alibek, *Biohazard* (Random House, 2008).

22 Alexander Shlyakhter and Richard Wilson, "Chernobyl: the inevitable results of secrecy," *Public Understanding of Science* 1.3 (1992): 251.

Chapter 2

23 "Man Angry at IRS Crashes Plane into Office," *CBS News*, February 18, 2010, https://www.cbsnews.com/news/man-angry-at-irs-crashes-plane-into-office/.

24 Spc. Linsey Williams 1st Armored Brigade Combat Team, "Minnesota National Guard commemorates 150th anniversary of Gettysburg," National Guard, July 3, 2013, https://www.nationalguard.mil/News/Article/574739/minnesota-national-guard-commemorates-150th-anniversary-of-gettysburg/.

25 United States Army Indirect Fire Infantryman Job Description, 2022, http://asktop.net/wp/download/10/STP7_11C14.pdf.

Chapter 3

26 "CENTRAL AMERICA WAR (1979-1992)," *USContraWar*, https://uscontrawar.com.

27 FORT BENNING: U.S. Army Fort Benning and The Maneuver Center of Excellence, https://www.benning.army.mil/infantry/198th/2-19/.

28 USAF Academy Legal Office, "STATUS OF FORCES AGREEMENT (SOFA)," https://www.usafa.af.mil/Portals/21/documents/Leadership/JudgeAdvocate/SOFA. pdf?ver=2015-10-30-115236-060.

29 Army Institute for Professional Development, COMBAT LIFESAVER COURSE: STUDENT SELF-STUDY, Subcourse ISO87C, Edition C, http://www.shastadefense. com/IS0871-CALMS.pdf.

30 Max Blumenthal and Ben Norton, "Max Blumenthal drops by the largest US military base in Latin America," *The Grayzone*, July 20, 2019, https://thegrayzone.com/2019/07/20/ max-blumenthal-palmerola-air-base-honduras/.

31 "MILITARY & TERRORIST ATTACKS IN HONDURAS," *USContraWar*, http:// uscontrawar.com/military-terrorist-attacks-in-honduras/.

32 "2004 Haitian coup d'état," Wikipedia, last modified October 11, 2022, https:// en.wikipedia.org/wiki/2004_Haitian_coup_d%27%C3%A9tat.

33 "Current Operations," Joint Task Force-Bravo, October 25, 2021, https://www.jtfb. southcom.mil/About-Us/Fact-Sheets/Display/Article/1860359/current-operations/.

34 "What is a SCIF?" SCIF Global, https://scifglobal.com/scif-definition-what-is-a-scif/.

35 Diana Roy, "China's Growing Influence in Latin America," Council on Foreign Relations, April 12, 2022, https://www.cfr.org/backgrounder/china-influence-latin-america-argentina-brazil-venezuela-security-energy-bri.

36 Britannica, T. Editors of Encyclopaedia. "FARC." Encyclopedia Britannica, October 14, 2022. https://www.britannica.com/topic/FARC.

37 "Stop-loss policy," Wikipedia, last modified January 13, 2022, https://en.wikipedia.org/ wiki/Stop-loss_policy.

Chapter 4

38 "Camp Ashraf," Wikipedia, last modified August 28, 2022, https://en.wikipedia.org/ wiki/Camp_Ashraf.

39 "…PMOI/MEK," U.S. Committee for Camp Ashraf Residents, https://www.usccar .org/about/pmoimek/.

40 "LSA Anaconda," *Global Security*, https://www.globalsecurity.org/military/world/iraq/ lsa-anaconda.htm.

Chapter 5

41 Mark Thompson, "America's Medicated Army," *Time*, June 5, 2008, https:// jeffreydachmd.com/wp-content/uploads/2015/03/TIME-Americas-Medicated-Army-Mark-Thompson-2008.pdf .

42 US Department of Veterans Affairs National Center for PTSD, "Iraq Clinician Treatment Guide," Treatment of the Returning Iraq War Veteran, https://www. globalsecurity.org/military/library/report/2004/Chapter_IV.pdf.

Chapter 6

43 "Contracting Officer's Technical Representative," Wikipedia, last modified June 14, 2021, https://en.wikipedia.org/wiki/Contracting_Officer%27s_Technical_Representative.

44 LinkedIn, "Thad Strom, PhD," https://www.linkedin.com/in/thad-strom-Ph.D.-abpp-89793951.

Chapter 7

45 "About DARPA," Defense Advanced Research Projects Agency, https://www.darpa.mil/about-us/about-darpa.
46 Ruth D. Peterson and Lauren J. Krivo, "National Neighborhood Crime Study (NNCS), 2000 (ICPSR 27501)," *Resource Center for Minority Data*, May 5, 2010, https://www.icpsr.umich.edu/web/RCMD/studies/27501.
47 "Protecting the global food supply through research, education, and the delivery of innovative solutions," University of Minnesota Food Protection and Defense Institute, https://foodprotection.umn.edu.
48 "National Infrastructure Protection Plan Agriculture and Food Sector," Department of Homeland Security, https://www.dhs.gov/xlibrary/assets/nipp_snapshot_agriculture.pdf.
49 "Government Coordinating Councils," Cybersecurity & Infrastructure Security Agency, https://www.cisa.gov/government-coordinating-councils.
50 Ibid.
51 "Master Government List of Federally Funded R&D Centers," *National Science Foundation*, February 2022, https://www.nsf.gov/statistics/ffrdclist/.
52 "About Sandia," *Sandia National Laboratories*, https://www.sandia.gov/about/.
53 "Melamine and food safety in China," *The Lancet*, Volume 373, Issue 9661, January 31, 2009, https://www.thelancet.com/journals/lancet/article/PIIS0140-6736(09)60114-8/fulltext.
54 "2008 – China," *World Health Organization*, September 18, 2008, https://www.who.int/emergencies/disease-outbreak-news/item/2008_09_19-en.
55 "The BioWatch Program Factsheet," Department of Homeland Security, June 6, 2022, https://www.dhs.gov/publication/biowatch-program-factsheet.

Chapter 9

56 "Our History," *Defense Threat Reduction Agency*, https://www.dtra.mil/About/DTRA-History/.
57 AG Huff, et al., "Biosurveillance: a systematic review of global infectious disease surveillance systems from 1900 to 2016," *Rev Sci Tech*, National Library of Medicine, August 2017, https://pubmed.ncbi.nlm.nih.gov/30152467/.
58 "Understanding the Risk of Bat Coronavirus Emergence," *The Intercept*, September 8, 2021, https://theintercept.com/document/2021/09/08/understanding-the-risk-of-bat-coronavirus-emergence/.https://theintercept.com/document/2021/09/08/understanding-the-risk-of-bat-coronavirus-emergence/.
59 "The Heilmeier Catechism," DARPA, https://www.darpa.mil/work-with-us/heilmeier-catechism.
60 "Doing Diligence to Assess the Risks and Benefits of Life Sciences Gain-of-Function Research," National Archives and Records Administration, Obama White House archives, October 17, 2014, https://obamawhitehouse.archives.gov/blog/2014/10/17/doing-diligence-assess-risks-and-benefits-life-sciences-gain-function-research.

61 "What Is Genomic Surveillance?" *Centers for Disease Control and Prevention*, January 24, 2022, https://www.cdc.gov/coronavirus/2019-ncov/variants/genomic-surveillance. html.

62 Eliot Marshall and Michael Price, "U.S. Supreme Court Strikes Down Human Gene Patents," *Science*, June 13, 2013, https://www.science.org/content/article/us-supreme-court-strikes-down-human-gene-patents.

63 Oversight Committee Republicans (@GOPoversight), "July 28th NIH says 'no NIAID funding was approved for Gain of Function research at the WIV,'" Twitter, October 20, 2021, 5:17 p.m., https://twitter.com/GOPoversight/status/1450934193177903105/ photo/1.

64 Sharon Lerner, Mara Hvistendahl, Maia Hibbett, "NIH DOCUMENTS PROVIDE NEW EVIDENCE U.S. FUNDED GAIN-OF-FUNCTION RESEARCH IN WUHAN," *The Intercept*, September 9, 2021, https://theintercept.com/2021/09/09/ covid-origins-gain-of-function-research/.

65 Science and Technology Policy Office, "United States Government Policy for Institutional Oversight of Life Sciences Dual Use Research of Concern," *Federal Register*, February 2, 2013, https://www.federalregister.gov/documents/2013/02/22/2013-04127/united-states-government-policy-for-institutional-oversight-of-life-sciences-dual-use-research-of#h-4.

66 US Government Science, Safety, Security, "United States Government Policy for Institutional Oversight of Life Sciences Dual Use Research of Concern," September 24, 2014, https://www.phe.gov/s3/dualuse/Documents/durc-policy.pdf.

Chapter 10

67 Mike Smith, PhD, "CRISPR," *National Human Genome Research Institute*, October 17, 2022, https://www.genome.gov/genetics-glossary/CRISPR.

68 "PREDICT (USAID)," Wikipedia, last modified July 11, 2022, https://en.wikipedia. org/wiki/PREDICT_(USAID).

69 Ibid.

Chapter 11

70 Andrew G. Huff, et al.,"Evaluation and Verification of the Global Rapid Identification of Threats System for Infectious Diseases in Textual Data Sources," *Interdisciplinary Perspectives on Infectious Diseases*, vol. 2016, Article ID 5080746, September 6, 2016, https://doi.org/10.1155/2016/5080746.

71 Andrew G. Huff, et al., "How resilient is the United States' food system to pandemics?" *Journal of Environmental Studies and Sciences*, 5, 337–347 (2015) https://doi. org/10.1007/s13412-015-0275-3.

72 Alex Gangitano, "Trump uses Defense Production Act to order meat processing plants to stay open," *The Hill*, April 28, 2020, https://thehill.com/homenews/ administration/495175-trump-uses-defense-production-act-to-order-meat-processing-plants-to/.

73 Donald G. McNeil Jr, "Starvation Timetable in a Pandemic," *New York Times*, June 19, 2015, https://www.nytimes.com/2015/06/23/health/starvation-timetable-in-a -pandemic.html.

74 A.G. Huff and T. Allen, "Mantle: A free and multilingual software for one health biosurveillance & research," *International Journal of Infectious Diseases*, Volume 45, Supplement 1, April 1, 2016, t https://doi.org/10.1016/j.ijid.2016.02.656.

75 Andrew Huff, et al., "FLIRT-ing with Zika: A Web Application to Predict the Movement of Infected Travelers Validated Against the Current Zika Virus Epidemic," *PLOS Currents*, June 10, 2016, https://www.ncbi.nlm.nih.gov/pmc/articles/PMC4922883/.

76 Kamala Kelker, "Zika virus: Texas and Florida airports at highest risk to receive infected travelers," *The Guardian*, February 5, 2016, https://www.theguardian.com/world/2016/feb/05/zika-virus-florida-texas-airports-risk-infection.

77 James S. Hodges, "Six (Or So) Things You Can Do with a Bad Model," *Operations Research*, Vol. 39, No. 3, June 1, 1991, https://pubsonline.informs.org/doi/pdf/10.1287/opre.39.3.355.

Chapter 12

78 Massachusetts Corporations Division, State Filing, EcoHealth Alliance, # 311726494, https://corp.sec.state.ma.us/CorpWeb/CorpSearch/CorpSummary.aspx?sysvalue=WpX wbLvwlMlEiHamIYWcb1G607Vt6JamJvuboLK49gY-.

79 "There are more than 1 million viruses that we know absolutely nothing about," *EcoHealth Alliance*, January 2017, https://www.ecohealthalliance.org/2017/01/1-million-viruses-know-absolutely-nothing.

80 Natalie Winters, "REVEALED: Google & USAID Funded Wuhan Collaborator Peter Daszak's Virus Experiments for Over A Decade," *The National Pulse*, June 19, 2021, https://thenationalpulse.com/2021/06/19/google-funded-wuhan-linked-ecohealth-research/.

81 Half-Day Meeting of the Board of Directors, EcoHealth Alliance, December 15, 2016, https://downloads.vanityfair.com/ecohealth-alliance/board-meeting-minutes.pdf.

82 Board of Directors, EcoHealth Alliance, https://www.ecohealthalliance.org/board-of-directors.

83 Samuel Chamberlain, "Pentagon gave millions to EcoHealth Alliance for weapons research program," *New York Post*, July 1, 2021, https://nypost.com/2021/07/01/pentagon-gave-millions-to-ecohealth-alliance-for-wuhan-lab/.

Chapter 13

84 "What is PREDICT?" *US Aid from the American People*, 2020, https://p2.predict.global/predict-project.

85 "Latest News," *Drastic Research*, https://drasticresearch.org.

86 Dennis A Benson et al., "GenBank." *Nucleic acids research* vol. 41, Database issue (2013): D36-42. https://www.ncbi.nlm.nih.gov/genbank/.

87 https://blast.ncbi.nlm.nih.gov/Blast.cgi.

88 "PREDICT Emerging Pandemic Threats Project," University of California Davis, *US Aid from the American People*, January24, 2022, https://data.usaid.gov/Global-Health-Security-in-Development-GHSD-/PREDICT-Emerging-Pandemic-Threats-Project/tqea-hwmr

89 https://github.com/jbloom/eidith-1.

90 Richard H. Ebright (@R_H_Ebright), "EcoHealth staff did not always adhere to their own safety requirements," Twitter, September 28, 2022, 4:58 p.m., https://twitter.com/R_H_Ebright/status/1575228366805815296.

91 "Bradley Fighting Vehicle," Wikipedia, last modified October 11, 2022, https://en.wikipedia.org/wiki/Bradley_Fighting_Vehicle.

92 Robert Roskoski Jr. and Tristram G. Parslow, "Ranking Tables of NIH Funding to US Medical Schools in 2020," *Blue Ridge Institute for Medical Research*, February 9, 2021, http://www.brimr.org/NIH_Awards/2020/default.htm.

93 Jack Posobiec, "VIDEO: Top Gain of Function Scientist Ralph Baric Admitted Viruses Can Be Lab Engineered 'Without a Trace'," *Human Events*, July 12, 2021, https://humanevents.com/2021/07/12/video-top-gain-of-function-scientist-ralph-baric-admitted-viruses-can-be-lab-engineered-without-a-trace/.

94 Kristopher M. Curtis, Boyd Yount, Ralph S. Baric, "Methods for producing recombinant coronavirus," US Patent number 7,279,327 B2, October 9, 2007, https://patents.google.com/patent/US7279327B2/en.

95 Vineet D Menachery, et al., "A SARS-like cluster of circulating bat coronaviruses shows potential for human emergence," *Nature Medicine*, November 9, 2015, https://doi.org/10.1038/nm. c.

96 Chris Pandolfo, "'An actual scandal': Report details how Wuhan lab and EcoHealth Alliance partnered to engineer coronaviruses in possibly unsafe conditions," *Blaze Media*, June 29, 2021, https://www.theblaze.com/news/wuhan-lab-ecohealth-unsafe-conditions.

97 Rowan Jacobsen, "Inside the risky bat-virus engineering that links America to Wuhan," *MIT Technology Review*, June 29, 2021, https://www.technologyreview.com/2021/06/29/1027290/gain-of-function-risky-bat-virus-engineering-links-america-to-wuhan/.

98 Sonia Shah, "History and Future of Pandemics," C-SPAN video, February 23, 2016, 1:35:06, https://www.c-span.org/video/?404875-1/pandemics#.

99 Peter Daszak (@PeterDaszak), "Not true - we've made great progress with bat SARS-related CoVs, ID'ing >50 novel strains," Twitter, November 21, 2019, 4:42 p.m., https://twitter.com/PeterDaszak/status/1197631383470034951.

Chapter 14

100 Obituary of Harvey David Kasdan, Tribute Archive (Mamaroneck, New York), February 11, 2016, https://www.tributearchive.com/obituaries/765023/Harvey-David-Kasdan/wall.

101 Cheryl Pellerin, "DTRA Program Helps Nations Tackle Biological Threats," *DOD News*, US Department of Defense, March 10, 2016, https://www.defense.gov/News/News-Stories/Article/Article/689971/dtra-program-helps-nations-tackle-biological-threats/.

102 US Embassy in Ukraine, "Biological Threat Reduction Program," https://ua.usembassy.gov/embassy/kyiv/sections-offices/defense-threat-reduction-office/biological-threat-reduction-program/.

103 Lance Brooks, "Cooperative Biological Engagement Program," US Defense Threat Reduction Agency, May 9, 2018, https://www.epa.gov/sites/default/files/2018-11/documents/decon_presentation_021.pdf.

104 "CBEP Ukraine Phase II — BTRIC TO 1," Black & Veatch, https://bv.com/projects/cbep-ukraine-phase-ii-btric-1.

105 Swarajya, "Ukraine Has Biological Research Facilities, Concerned Russian Forces May Seek To Gain Control," YouTube video, March 11, 2022, https://www.youtube.com/watch?v=ydSf57SRtcQ.

106 Senator Marco Rubio, "Vice Chairman Rubio Questions DNI Haines on Biolab Security," YouTube video, March 10, 2022, https://www.youtube.com/watch?v=2lUt6DMfrBg.

107 Steve Sweeney, "After months of denial, U.S. admits to running Ukraine biolabs," *People's World*, June 14, 2022, https://www.peoplesworld.org/article/after-months-of-denial-u-s-admits-to-running-ukraine-biolabs/.

108 "Munich Re & In-Q-Tel Select Metabiota to Gain Deeper Insights into Epidemic Risk and Global Preparedness for Infectious Diseases," IQT, August 22, 2017, https://www.iqt.org/news/munich-re-in-q-tel-select-metabiota-to-gain-deeper-insights-into-epidemic-risk-and-global-preparedness-for-infectious-diseases/.

109 Joe Hoft, "EXCLUSIVE: CEO of Metabiota, Nathan Wolfe, Connected to Hunter Biden, the NIH, the CDC and the World Economic Forum," *Gateway Pundit*, March 12, 2022, https://www.thegatewaypundit.com/2022/03/exclusive-ceo-metabiota-nathan-wolfe-connected-hunter-biden-nih-cdc-world-economic-forum/.

110 "Flashback: Hunter Biden's investment firm led financing for key partners of Wuhan lab," *World Tribune*, March 1, 2022, https://www.worldtribune.com/flashback-hunter-bidens-investment-firm-led-financing-for-key-partners-of-wuhan-lab/.

111 Srivats Lakshman, "What is Metabiota? Emails reveal Hunter Biden got millions for DoD contractor specializing in infectious diseases," *MEAWW*, March 26,2022, https://meaww.com/metabiota-dod-contracts-hunter-biden-links-rstp-funding-russian-claims-laptop-emals.

112 Dilyana Gaytandzhieva, "Documents expose US biological experiments on allied soldiers in Ukraine and Georgia," *Arms Watch*, January 25, 2022, https://armswatch.com/documents-expose-us-biological-experiments-on-allied-soldiers-in-ukraine-and-georgia/.

113 Dilyana Gaytandzhieva, "Georgia cover-up of deaths in $3.3 billion pharmaceutical project: documents," *Arms Watch*, September 9, 2020, https://armswatch.com/georgia-cover-up-of-deaths-in-3-3-billion-pharmaceutical-project-documents/.

114 Dilyana Gaytandzhieva, "Pentagon Biolaboratories - Investigative Documentary," YouTube video, September 21, 2018, https://www.youtube.com/watch?v=_8hQi2Zv1L0.

115 S. Nundy & C.M. Gulhati, "A new colonialism?—Conducting clinical trials in India," *New England journal of medicine*, 352(16), (2005)1633-1636.

116 "'Explosive' Growth in Foreign Drug Testing Raises Ethical Questions," *PBS News Hour*, August 23, 2011, https://www.pbs.org/newshour/health/sending-us-drug-research-overseas.

117 Thomas Pogge, "Testing our drugs on the poor abroad," *ResearchGate*, August 2008, https://www.researchgate.net/publication/287001271_Testing_our_drugs_on_the_poor_abroad.

118 Tessa Richards, "Conduct of drug trials in poor countries must improve," *BMJ*, June 23, 2005, https://www.ncbi.nlm.nih.gov/pmc/articles/PMC558481/.

119 Stephanie Kelly, "Testing Drugs on the Developing World," *The Atlantic*, February 27, 2013, https://www.theatlantic.com/health/archive/2013/02/testing-drugs-on-the-developing-world/273329/.

120 Dilyana Gaytandzhieva, "Documents expose US biological experiments on allied soldiers in Ukraine and Georgia," January 25, 2022, https://armswatch.com/documents-expose -us-biological-experiments-on-allied-soldiers-in-ukraine-and-georgia/.

121 Samia Arshad, et al. "Treatment with hydroxychloroquine, azithromycin, and combination in patients hospitalized with COVID-19," *International journal of infectious diseases:* IJID : official publication of the International Society for Infectious Diseases vol. 97 (2020): 396-403. doi:10.1016/j.ijid.2020.06.099.

Chapter 15

122 Truman Leadership Quote, *Leadership Weblog,* June 5, 2011, https://blogs.harvard.edu/ leadership/truman-leadership-quote/.

123 Andrew G. Huff, "Enhancing Food Defense: Risk Managers' Perceptions, Criticality Assessments, and a Novel Method for Objectively Determining Food Systems' Criticality," (PhD diss., University of Minnesota, 2014), https://conservancy.umn.edu/ bitstream/handle/11299/163766/Huff_umn_0130E_14846.pdf?sequence=1.

124 Robert W. Pinner, MD, et al., "Trends in Infectious Diseases Mortality in the United States," *JAMA,* January 17, 1996, https://jamanetwork.com/journals/jama/article-abstract/394322.

125 Max Roser, et al., "World Population Growth," *Our World in Data,* May 2019, https:// ourworldindata.org/world-population-growth.

126 Stephen Endicott and Edward Hagerman, *The United States and Biological Warfare: Secrets from the Early Cold War and Korea* (Indiana University Press, 1998).

127 Edward Drea, et al., *Researching Japanese War Crimes Records* (Washington DC: National Archives, 2006), https://www.archives.gov/files/iwg/japanese-war-crimes/introductory-essays.pdf.

128 Roser, op. cit.

129 Friedrich Frischknecht, "The history of biological warfare," *EMBO Reports* (2003)4:S47-S52, May 9, 2003, doi: 10.1038/sj.embor.embor849 /.

130 Jeffrey K. Smart M.A., *History of Chemical and Biological Warfare,* Chapter 2, https:// studylib.net/doc/11661349/chapter-2-history-of-chemical-and-biological-warfare--an-...

131 Michelle Rozo and Gigi Kwik Gronvall, "The Reemergent 1977 H1N1 Strain and the Gain-of-Function Debate," *mBio,* August 18, 2015, https://doi.org/10.1128/ mBio.01013-15.

132 George C. Wilson, "Army Conducted 239 Secret, Open-Air Germ Warfare Tests," *Washington Post,* March 9, 1977, https://www.washingtonpost.com/archive/ politics/1977/03/09/army-conducted-239-secret-open-air-germ-warfare-tests/ b17e5ee7-3006-4152-acf3-0ad163e17a22/.

133 A T. Tu, "Aum Shinrikyo's Chemical and Biological Weapons: More Than Sarin." *Forensic science review* vol. 26,2 (2014): 115-20, https://pubmed.ncbi.nlm.nih. gov/26227027/.

134 "CBMs," Biological Weapons Convention, 2011, http://bwc1972.org/home/the-biological-weapons-convention/about-the-bwc/text-of-the-biological-weapons-convention-2/cbms/.

135 Biological Weapons Convention, United Nations Office for Disarmament Affairs, https://www.un.org/disarmament/biological-weapons/.

136 U.S. Congress, *Uniting and Strengthening America by Providing Appropriate Tools Required to Intercept and Obstruct Terrorism (USA Patriot Act) Act of 2001*, HR 3162, 107th Congress, October 26, 2001, https://www.congress.gov/107/plaws/publ56/PLAW-107publ56.pdf.

137 U. S. Congress, *Public Health Security and Bioterrorism Preparedness and Response Act of 2002*, HR 3448, 107th Congress, June 12, 2002, https://www.congress.gov/107/plaws/publ188/PLAW-107publ188.pdf.

138 National Research Council, *Biotechnology Research in an Age of Terrorism* (Washington, DC: The National Academies Press, 2004), Chapter 1, https://doi.org/10.17226/10827.

139 NIH Grants Policy Statement, *National Institutes of Health*, December 2021, 15.1 General, https://grants.nih.gov/grants/policy/nihgps/HTML5/section_15/15.1_general.htm.

140 National Research Council, *Biotechnology Research in an Age of Terrorism*.

141 Robert Roos, "NSABB member says officials stacked deck for board's H5N1 decision," *Center for Infectious Disease Research and Policy*, April 13, 2012, https://www.cidrap.umn.edu/news-perspective/2012/04/nsabb-member-says-officials-stacked-deck-boards-h5n1-decision.

142 Ibid.

Chapter 16

143 Li Mingyuan, et. al., "The nano delivery systems and applications of mRNA," *European Journal of Medicinal Chemistry* Volume 227, 113910, January 5, 2022, https://www.ncbi.nlm.nih.gov/pmc/articles/PMC8497955/#bib7.

144 Elie Dolgin, "The tangled history of mRNA vaccines," *Nature*, September 14, 2021, https://www.nature.com/articles/d41586-021-02483-w.

145 "Outlook on the mRNA Global Market and Therapeutics to 2026," *Globe Newswire*, October 6, 202, https://www.globenewswire.com/news-release/2021/10/06/2309400/28124/en/Outlook-on-the-mRNA-Global-Market-and-Therapeutics-to-2026.html.

146 Chris Beyrer, "The Long History of mRNA Vaccines," *Johns Hopkins*, October 6, 2021, https://publichealth.jhu.edu/2021/the-long-history-of-mrna-vaccines.

147 Ralph S. Baric, et al., "High Recombination and Mutation Rates in Mouse Hepatitis Virus Suggest that Coronaviruses may be Potentially Important Emerging Viruses," In: Talbot, P.J., Levy, G.A. (eds) *Corona- and Related Viruses. Advances in Experimental Medicine and Biology*, vol 380. Springer, Boston, MA. https://doi.org/10.1007/978-1-4615-1899-0_91.

148 K. M. Curtis, B. Yount, R. S. Baric, "Heterologous Gene Expression from Transmissible Gastroenteritis Virus Replicon Particles," Carolina Digital Repository, 2002, https://cdr.lib.unc.edu/concern/articles/tb09jf50d?locale=en.

149 U.S. Department of State Archive, "BioIndustry Initiative," September 16, 2003, https://2001-2009.state.gov/t/isn/rls/fs/24242.htm.

150 Caren Black, "Another Failure of Imagination?" *Titanic Lifeboat Academy*, November 21, 2020, https://www.titaniclifeboatacademy.org/34-teolawki-times/1297-another-failure-of-imagination.

151 Vaccine & Immunotherapy Center, bio of Dr. Michael V. Callahan, https://advancingcures.org/vic-management-michael-callahan/.

152 Raul Diego, "DARPA's Man in Wuhan," *Silicon Icarus*, July 31, 2020, https://siliconicarus.org/2020/07/31/darpas-man-in-wuhan/.

153 "'My name is Spartacus': COVID-19 Deep Dive Part III: Criminal Conspiracy," *The Expose*, March 21, 2022, https://expose-news.com/2022/03/21/spartacus-covid-19-deep-dive-part-iii-criminal-conspiracy/.

154 Roy M. Anderson, et al., "Epidemiology, transmission dynamics and control of SARS: the 2002–2003 epidemic," *The Royal Society*, June 15, 2004, https://www.ncbi.nlm.nih.gov/pmc/articles/PMC1693389/pdf/15306395.pdf.

155 House of Representatives, Committee of Homeland Security, *Lessons from the Offensive U.S. and Russian Biological Weapons Programs*, 109th Congress, July 13, 2005, https://irp.fas.org/congress/2005_hr/bioterror.html.

156 "'My name is Spartacus': COVID-19 Deep Dive Part III: Criminal Conspiracy," *The Expose*, March 21, 2022, https://expose-news.com/2022/03/21/spartacus-covid-19-deep-dive-part-iii-criminal-conspiracy/.

157 S.313 - Nunn-Lugar Cooperative Threat Reduction Act of 2005, 109th Congress (2005-2006) https://www.congress.gov/bill/109th-congress/senate-bill/313/text.

158 "Fact Sheet: The Nunn-Lugar Cooperative Threat Reduction Program," *Center for Arms Control*, March 29, 2022, https://armscontrolcenter.org/fact-sheet-the-nunn-lugar-cooperative-threat-reduction-program-2/.

159 Rose Gottemoeller, "Science diplomacy: The essential interdisciplinary approach," *Bulletin of the Atomic Scientists*, December 7, 2020, https://thebulletin.org/premium/2020-12/science-diplomacy-the-essential-interdisciplinary-approach/.

160 Defense Advanced Research Projects Agency, "Accelerated Manufacturing of Pharmaceuticals Teaming and Information Session," May 10, 2006, https://web.archive.org/web/20201205052621/https://fbo.gov.surf/FBO/Solicitation/SN06-26.

161 Peter Daszak, "Risk of Viral Emergence from Bats," *EcoHealth Alliance*, https://grantome.com/grant/NIH/R01-AI079231-02.

162 Ibid.

163 Defense Advanced Research Projects Agency (DARPA) and University of Pittsburgh Medical Center (UPMC), "Ensuring Biologics Advanced Development and Manufacturing Capability for the United States Government: A Summary of Key Findings and Conclusions," July 2007-March 2009, https://apps.dtic.mil/dtic/tr/fulltext/u2/a506569.pdf.

164 Recorded Webinar: Operation Warp Speed COVID-19 Vaccine Distribution Strategy Briefing, US Department of Health and Human Services, *Public Health Emergency*, October 2020, https://www.phe.gov/Preparedness/planning/cip/Pages/OWS-Vaccine-Briefing.aspx.

165 Sixth Amendment to Declaration Under the Public Readiness and Emergency Preparedness Act for Medical Countermeasures Against COVID–19 , Department of Health and Human Services, *Public Health Emergency*, February 10, 2021, https://www.phe.gov/Preparedness/legal/prepact/Pages/COVID-Amendment-6.aspx.

166 "Reducing Pandemic Risk, Promoting Global Health," *USAID, PREDICT*, https://www.usaid.gov/sites/default/files/documents/1864/predict-global-flyer-508.pdf.

167 PREDICT Program, USAID Emerging Pandemic Threats, UC Davis, 2009-2020, https://ohi.vetmed.ucdavis.edu/programs-projects/predict-project.

168 Jane Qiu, "How China's 'Bat Woman' Hunted Down Viruses from SARS to the New Coronavirus," *Scientific American*, June 1, 2020, https://www.scientificamerican.

com/article/how-chinas-bat-woman-hunted-down-viruses-from-sars-to-the-new-coronavirus1/.

169 PREDICT Program, USAID Emerging Pandemic Threats.

170 Biography: Dr. David R. Franz, Harvard T.H. Chan, School of Public Health, https://theforum.sph.harvard.edu/expert-participants/david-r-franz/.

171 Sam Husseini, "Militarized Pandemic Science: Why is the Pentagon Funding the EcoHealth Alliance?" *CounterPunch*, December 21, 2020 https://www.counterpunch.org/2020/12/21/militarized-pandemic-science-why-is-the-pentagon-funding-the-ecohealth-alliance/.

172 Biography: Robert Kadlec, M.D., Assistant Secretary for Preparedness and Response, US Department of Health and Human Services, https://www.phe.gov/newsroom/bio/Documents/kadlec-bio-print.pdf.

173 National Science Advisory Board for BioSecurity Meeting Minutes, October 22, 2014 https://osp.od.nih.gov/wp-content/uploads/NSABB_Meeting_Minutes-Oct_2014.pdf.

174 Dr. Ah Kahn Syed, "Absolute proof: The Gp-120 sequences prove beyond all doubt that "COVID-19" was man-made," *Arkmedic's Blog*, April 10, https://arkmedic.substack.com/p/absolute-proof-the-gp-120-sequences.

175 Dr. Ah Kahn Syed, "How to BLAST your way to the truth about the origins of COVID-19," *Arkmedic's Blog*, December 28, 2021, https://arkmedic.substack.com/p/how-to-blast-your-way-to-the-truth.

176 Steven Carl Quay MD PhD, "A Bayesian analysis concludes beyond a reasonable doubt that SARS-CoV-2 is not a natural zoonosis but instead is laboratory derived," *Zenodo*, January 29, 2021, https://zenodo.org/record/4477081#.Y0hSFC-B0Us.

177 Steven Carl Quay MD PhD, "An Introduction to a Bayesian Analysis of the Laboratory Origin of SARS-CoV-2," *Zenodo*, January 29, 2021, https://zenodo.org/record/4477212#.Y0hSFS-B0Us.

178 "Wuhan seafood market pneumonia virus isolate Wuhan-Hu-1, complete genome," *National Library of Medicine*, https://www.ncbi.nlm.nih.gov/nuccore/NC_045512.1.

179 K. Balamurali et al, "MSH3 Homology and Potential Recombination Link to SARS-CoV-2 Furin Cleavage Site," *Frontiers in Virology*, February 21, 2022, https://www.frontiersin.org/articles/10.3389/fviro.2022.834808/full.

180 Stephen V. Su, et al., "DC-SIGN Binds to HIV-1 Glycoprotein 120 in a Distinct but Overlapping Fashion Compared with ICAM-2 and ICAM-3," *Journal of Biological Chemistry*, Volume 279, Issue 18, 2004, Pages 19122-19132, https://www.sciencedirect.com/science/article/pii/S0021925819754751.

181 Peak Prosperity, "Fauci's Dishonesty and Co-Conspirators Revealed!" Youtube video, June 4, 2021, https://www.youtube.com/watch?v=DNxoVFZwMYw.

182 Josh Rogin, "State Department cables warned of safety issues at Wuhan lab studying bat coronaviruses," *Washington Post*, April 14, 2020, https://www.washingtonpost.com/opinions/2020/04/14/state-department-cables-warned-safety-issues-wuhan-lab-studying-bat-coronaviruses/.

183 "FOLLOWING 2018 STATE DEPARTMENT WARNINGS OF SAFETY ISSUES AT WUHAN LAB STUDYING CORONAVIRUSES, MURPHY, MARKEY PRESS POMPEO FOR ANSWERS ON TRUMP ADMINISTRATION'S RESPONSE," Chris Murphy, April 28, 2020 https://www.murphy.senate.gov/newsroom/press-releases/following-2018-state-department-warnings-of-safety-issues-at-wuhan-lab-studying

-coronaviruses-murphy-markey-press-pompeo-for-answers-on-trump-administrations
-response-.

[184] Rogin, op. cit.

[185] Noah Higgins-Dunn, "Biden coronavirus advisor Osterholm says U.S. is 'about to enter Covid hell'," *CNBC*, November 9, 2020, https://www.cnbc.com/2020/11/09/president-elect-biden-coronavirus-advisor-osterholm-says-us-is-about-to-enter-covid-hell.html.

[186] Vineet D Menachery, et al., "A SARS-like cluster of circulating bat coronaviruses shows potential for human emergence," *Naturemedicine*, November 9, 2015, https://www.nature.com/articles/nm.3985.

[187] Renz Law, LLC, letter of September 12, 2022, https://tomrenz.substack.com/api/v1/file/42e49d09-33f1-4548-a7f2-7f87037f1d00.pdf.

[188] Vineet D. Menachery, et al., "SARS-like WIV1-CoV poised for human emergence," *PNAS*, March 2016, https://www.pnas.org/doi/abs/10.1073/pnas.1517719113.

[189] Global Virome Project, https://www.globalviromeproject.org.

[190] "Pandemics of Yesterday, Today and Tomorrow: Are We Always Forgetting a Piece of the Puzzle?" Institute of Health and Society webinar, June 4, 2020, https://www.med.uio.no/helsam/english/research/centres/global-health/news-and-events/events/2020/pandemics-of-yesterday-today---tomorrow.html.

[191] Shannon Murray, "Public comments on the WHO Scientific Advisory Group for the Origins of Novel Pathogens (SAGO) members," U.S. RIGHT TO KNOW, October 26, 2021, https://usrtk.org/covid-19-origins/public-comments-sago/.

[192] Matt Windsor, Master of Disaster, *UAB Magazine*, Spring/Summer 2021, https://www.uab.edu/uabmagazine/2013/february/callahan.

[193] Definitive Contract HR001111C0094, *GovTribe*, September 16, 2020, https://govtribe.com/award/federal-contract-award/definitive-contract-hr001111c0094.

[194] "DARPA's Man in Wuhan," *Unlimited Hangout*, July 31, 2020, https://unlimitedhangout.com/2020/07/investigative-reports/darpas-man-in-wuhan/.

[195] Raphael Satter, "Issues with false positives," *Associated Press*, July 17, 2014, https://www.documentcloud.org/documents/2752219-20140717-Report-on-Ebola-Response-by-Viral.html.

[196] Dan Edge, Sasha Joelle Achilli, "Outbreak," documentary transcript, *PBS Frontline*, September 2014, https://www.pbs.org/wgbh/frontline/documentary/outbreak/transcript/.

[197] Diego, op. cit.

[198] Whitney Webb, "Engineering Contagions - a 4 part Series by Whitney Webb, *Informed Choice Australia*, June 10, 2021, https://www.informedchoiceaustralia.com/post/engineering-contagions-a-4-part-series-by-whitney-webb.

[199] "The U.S. Counselor visited Wuhan Institute of Virology, CAS," March 4, 2018, https://web.archive.org/web/20180511170249/http:/english.whiov.cas.cn/Exchange2016/Foreign_Visits/201804/t20180403_191334.html.

[200] Wuhan Institute of Virology Partnerships, https://web.archive.org/web/20210209172907/http://english.whiov.cas.cn/International_Cooperation2016/Partnerships/.

[201] "Unclassified Summary of Assessment on COVID-19 Origins," National Intelligence Council, https://www.dni.gov/files/ODNI/documents/assessments/Unclassified-Summary-of-Assessment-on-COVID-19-Origins.pdf.

[202] Suchandrima Bhowmik, "Italian study finds SARS-CoV-2 in clinical samples collected before December 2019," *News Medical*, August 30, 2022, https://www.news-medical.

net/news/20220830/Italian-study-finds-SARS-CoV-2-in-clinical-samples-collected-before-December-2019.aspx.

203 Giovanni Apolone, et al. "Unexpected detection of SARS-CoV-2 antibodies in the prepandemic period in Italy," *Tumori* vol. 107,5 (2021): 446-451. doi:10.1177/0300891620974755.

204 Carlo Capalbo, et al., "No Evidence of SARS-CoV-2 Circulation in Rome (Italy) during the Pre-Pandemic Period: Results of a Retrospective Surveillance," *International journal of environmental research and public health* vol. 17,22 8461. 16 Nov. 2020, doi:10.3390/ijerph17228461.

205 Apolone, op cit.

206 Alessia Lai, Stefano, et al., "Evidence of SARS-CoV-2 Antibodies and RNA on Autopsy Cases in the Pre-Pandemic Period in Milan (Italy)," *Frontiers*, June 15, 2022, https://www.frontiersin.org/articles/10.3389/fmicb.2022.886317/full.

207 Scott LaFee, "Novel Coronavirus Circulated Undetected Months before First COVID-19 Cases in Wuhan, China," *UC San Diego Health*, March 18, 2021, https://health.ucsd.edu/news/releases/Pages/2021-03-18-novel-coronavirus-circulated-undetected-months-before-first-covid-19-cases-in-wuhan-china.aspx.

208 "COVID-19 might have started to spread in September 2019 in the United States: study," XinhuaNet, http://www.news.cn/english/2021-09/23/c_1310203268.htm.

209 "Argus-A Global Bioevent Tracking System," FOIA Electronic Reading Room, June 22, 2015, https://www.cia.gov/readingroom/document/0001523123.

210 Dr. Joseph Mercola, "Leaked Confidentiality Agreement Shows Moderna, NIAID Filed Covid Vaccine Candidate in 2019," *Truth Based Media*, July 9, 2021, https://truthbasedmedia.com/2021/07/09/leaked-confidentiality-agreement-shows-moderna-niaid-filed-covid-vaccine-candidate-in-2019/.

211 "Confidential Documents reveal Moderna sent mRNA Coronavirus Vaccine Candidate to University Researchers weeks before emergence of Covid-19," *The Expose*, June 18, 2021, https://expose-news.com/2021/06/18/confidential-documents-reveal-moderna-sent-mrna-coronavirus-vaccine-candidate-to-university-researchers-weeks-before-emergence-of-covid-19/.

212 "CDC Museum COVID-19 Timeline," *Centers for Disease Control and Prevention*, https://www.cdc.gov/museum/timeline/covid19.html.

213 Harvard2theBigHouse, "Golden Silkworms in Pandora's Box," *Harvard to the Big House*, April 1, 2021, https://harvard2thebighouse.substack.com/p/understanding-covid-19-and-seasonal.

214 K. Sirotkin, D. Sirotkin, "Might SARS-CoV-2 Have Arisen via Serial Passage through an Animal Host or Cell Culture?: A potential explanation for much of the novel coronavirus' distinctive genome," *Bioessays*, 2020 https://www.ncbi.nlm.nih.gov/pmc/articles/PMC7435492/.

215 Philip W. Magness, James R. Harrigan, "Fauci, Emails, and Some Alleged Science," *AIER*, December 19, 2021, https://www.aier.org/article/fauci-emails-and-some-alleged-science/.

216 KanekoatheGreat, "NIH Director Asked Fauci to Do a "Devastating Published Take Down" of "Fringe" Harvard, Stanford, and Oxford Epidemiologist," *Kanekoa News*, December 18, 2021, https://kanekoa.substack.com/p/nih-director-asked-fauci-to-do-a.

217 Tirtha Chakraborty, Antonin de Fougerolles, Modified polynucleotides encoding septin-4, US Patent US9149506B2, October 6, 2015, https://patents.google.com/patent/US9149506B2/en.

218 Tirtha Chakraborty, Antonin de Fougerolles, Modified polynucleotides encoding granulysin, US Patent US9216205B2, December 12, 2015, https://patents.google.com/patent/US9216205B2/en.

219 Tirtha Chakraborty, Antonin de Fougerolles, Modified polynucleotides encoding SIAH E3 ubiquitin protein ligase 1, US Patent, US9255129B2, February 9, 2016, https://patents.google.com/patent/US9255129B2/en.

220 Tirtha Chakraborty, Antonin de Fougerolles, Modified polynucleotides encoding apoptosis inducing factor 1, US Patent US9301993B2, April 5, 2016, https://patents.google.com/patent/US9301993B2/en.

221 "PREP Act Immunity from Liability for COVID-19 Vaccinators," Public Health Emergency, April 13, 2021, https://www.phe.gov/emergency/events/COVID19/COVIDVaccinators/Pages/PREP-Act-Immunity-from-Liability-for-COVID-19-Vaccinators.aspx.

222 "Expanding Access to COVID 19 Therapeutics," Public Health Emergency, September 13, 2021, https://www.phe.gov/Preparedness/legal/prepact/Pages/PREPact-NinethAmendment.aspx.

223 United States Senate, S.2244 – PREP Act of 2021, 117th Congress, June 24, 2021, https://www.congress.gov/bill/117th-congress/senate-bill/2244/amendments?r=22.

224 Derek Knauss, "Dr. Fleming Testimony," video, *Prepare for Change*, July 10, 2022, https://prepareforchange.net/2022/07/10/dr-fleming-testimony/.

225 United States Senate, Testimony of Dr. McCullough, November 19, 2020, https://www.hsgac.senate.gov/imo/media/doc/Testimony-McCullough-2020-11-19.pdf.

226 "Heroic Testimony from Drs. McCullough, Risch, and Fareed,"*AAPS*, November 19, 2020, https://aapsonline.org/heroic-testimony-from-drs-mccullough-risch-and-fareed/.

227 Major Joe Murphy USMC, email, August 13, 2021, https://assets.ctfassets.net/syq3snmxclc9/2mVob3c1aDd8CNvVnyei6n/95af7dbfd2958d4c2b8494048b4889b5/JAG_Docs_pt1_Og_WATERMARK_OVER_Redacted.pdf.

228 "Possession, Use, and Transfer of Select Agents and Toxins," Part 73, Federal Register, March 18, 2005, https://www.ecfr.gov/current/title-42/chapter-I/subchapter-F/part-73.

229 Connor Boyd, "Scientists claim Covid virus contains tiny chunk of DNA that 'matches sequence patented by Moderna THREE YEARS before pandemic began'," *Daily Mail*, February 23, 2022, https://www.dailymail.co.uk/news/article-10542309/Fresh-lab-leak-fears-study-finds-genetic-code-Covids-spike-protein-linked-Moderna-patent.html.

230 Balamurali K. Ambati, et al., "MSH3 Homology and Potential Recombination Link to SARS-CoV-2 Furin Cleavage Site," *Frontiers in Virology*, February 21, 2022, https://www.frontiersin.org/articles/10.3389/fviro.2022.834808/full.

Chapter 17

231 "China Spins Tale That the U.S. Army Started the Coronavirus Epidemic," *New York Times*, March 3, 2020, https://www.nytimes.com/2020/03/13/world/asia/coronavirus-china-conspiracy-theory.html.

232 Joseph Menn, "Pro-China social media campaign hits new countries, blames U.S. for COVID," *Reuters*, September 9, 2021, https://www.reuters.com/world/pro-china-social-media-campaign-expands-new-countries-blames-us-covid-2021-09-08/.

233 "Wuhan lab leak theory: How Fort Detrick became a centre for Chinese conspiracies," *BBC News*, August 23, 2021, https://www.bbc.co.uk/news/world-us-canada-58273322.

234 James Palmer, "Beijing Knows Who to Blame for the Virus: America," *Foreign Policy*, March 2, 2020, https://foreignpolicy.com/2020/03/02/china-blames-united-states-coronavirus/.

235 Xiao Qiang, "The Road to Digital Unfreedom: President Xi's Surveillance State," *Journal of Democracy*, January 2019, https://www.journalofdemocracy.org/articles/the-road-to-digital-unfreedom-president-xis-surveillance-state/.

236 Paul Mozur, Muyi Xiao, and John Liu, "How China is Policing the Future," *New York Times*, June 25, 2022, https://www.nytimes.com/2022/06/25/technology/china-surveillance-police.html.

237 Neil Thomas, "Members Only: Recruitment Trends in the Chinese Communist Party," *Macro Polo*, July 15, 2020, https://macropolo.org/analysis/members-only-recruitment-trends-in-the-chinese-communist-party/.

238 R. W. McMorrow, "Membership in the Communist Party of China: Who is Being Admitted and How?" *JSTOR Daily*, December 19, 2015, https://daily.jstor.org/communist-party-of-china/.

239 https://www.fbi.gov/news/speeches/the-threat-posed-by-the-chinese-government-and-the-chinese-communist-party-to-the-economic-and-national-security-of-the-united-states.

240 Jeff Ferry, "Top 10 Cases of Chinese IP Theft," *Coalition for a Prosperous America*, May 1, 2018, https://prosperousamerica.org/top-ten-cases-of-chinese-ip-theft/.

241 Kevin R. Brock, "China Cheats—and We Let Them," *The Hill*, October 7, 2019, https://thehill.com/opinion/international/464586-china-cheats-and-we-let-them/.

242 Kor Kian Beng, "In China, Honesty is Not the Best Policy," *Straits Times*, February 28, 2013, https://www.straitstimes.com/asia/in-china-honesty-is-not-the-best-policy.

243 Borge Bakken and Jasmine Wang, "The changing forms of corruption in China," *Crime Law Soc Change*, April 27, 2021, https://link.springer.com/article/10.1007/s10611-021-09952-3.

244 Meredith Wadman, "Having SARS-CoV-2 once confers much greater immunity than a vaccine—but vaccination remains vital," *Science Insider*, August 26, 2021, https://www.science.org/content/article/having-sars-cov-2-once-confers-much-greater-immunity-vaccine-vaccination-remains-vital.

245 M. S. Diamond, T. D. Kanneganti, "Innate immunity: the first line of defense against SARS-CoV-2," *Nat Immuno*, February 1, 2022, https://doi.org/10.1038/s41590-021-01091-.

246 Jennifer Block, "Vaccinating people who have had covid-19: why doesn't natural immunity count in the US?" *BMJ*, September 13, 2021, https://www.bmj.com/content/374/bmj.n2101/rr-0.

247 "Everybody's Business: Strengthening International Cooperation in a More Interdependent World Report of the Global Redesign Initiative," *World Economic Forum*, 2010, https://www3.weforum.org/docs/WEF_GRI_EverybodysBusiness_Report_2010.pdf.

248 Covax, *World Health Organization*, https://www.who.int/initiatives/act-accelerator/covax.

249 José Manuel Barroso, "Intellectual Property and COVID-19 vaccines," *Vaccines Work*, August 3, 2021, https://www.gavi.org/vaccineswork/intellectual-property-and-covid-19-vaccines.

250 Philip Loft, "Waiving intellectual property rights for Covid-19 vaccines," Research Briefing Number 9417, House of Commons Library, July 13, 2022, https://researchbriefings.files.parliament.uk/documents/CBP-9417/CBP-9417.pdf.

251 "Director-General's opening remarks at the media briefing on COVID-19 – 7 May 2021," *World Health Organization*, May 7, 2021, https://www.who.int/director-general/speeches/detail/director-general-s-opening-remarks-at-the-media-briefing-on-covid-19-7-may-2021.

252 Alexander Zaitchik, "How Bill Gates Impeded Global Access to Covid Vaccines," *New Republic*, April 12, 2021, https://newrepublic.com/article/162000/bill-gates-impeded-global-access-covid-vaccines.

253 Amy Maxmen, "The radical plan for vaccine equity," *Nature*, July 13, 2022, https://www.nature.com/immersive/d41586-022-01898-3/index.html.

254 Ashley Rindsberg, "China 'began stockpiling PPE months before Covid outbreak'," *The Telegraph*, October 8, 2022, https://www.telegraph.co.uk/news/2022/10/08/china-began-stockpiling-ppe-months-covid-outbreak/.

255 Antony Ashkenaz, "Covid origin theory blown open as China stockpiled West's PPE 'months' before outbreak," *Express*, October 10, 2022, https://www.express.co.uk/news/science/1680769/covid-origin-mystery-deepens-china-ppe-stockpile-virus-outbreak-wuhan.

256 "China Began Stockpiling Personal Protective Equipment Months Before Covid Outbreak," *Adverse Reaction Report*, October 2022, https://adversereactionreport.com/news/china-began-stockpiling-personal-protective-equipment-months-before-covid-outbreak/.

257 "China began stockpiling PPE before the Covid epidemic," *Share Talk,* October 9, 2022, https://www.share-talk.com/china-began-stockpiling-ppe-before-the-covid-epidemic/.

Chapter 19

258 By Samuel Chamberlain, Mark Moore and Bruce Golding, "Fauci was warned that COVID-19 may have been 'engineered,' emails show," *New York Post*, June 2, 2021, https://nypost.com/2021/06/02/fauci-was-warned-that-covid-may-have-been-engineered-emails/.

259 Katherine Eban, "'This Shouldn't Happen': Inside the Virus-Hunting Nonprofit at the Center of the Lab-Leak Controversy," *Vanity Fair*, March 31, 2022, https://www.vanityfair.com/news/2022/03/the-virus-hunting-nonprofit-at-the-center-of-the-lab-leak-controversy.

260 K. G. Andersen, et al., "The proximal origin of SARS-CoV-2," *Nat Med* 26, 450–452 April, 2020, https://doi.org/10.1038/s41591-020-0820-9.

261 Charles Calisher, et al., "Statement in support of the scientists, public health professionals, and medical professionals of China combatting COVID-19," *The Lancet*, February 19, 2020, https://www.thelancet.com/journals/lancet/article/PIIS0140-6736(20)30418-9/fulltext.

262 https://reporter.nih.gov/search/tVFRjbzaKk-KQ9ylFELqZw/projects.

Chapter 20

263 "Domestic Approach to National Intelligence," Department of National Intelligence, https://www.dni.gov/files/documents/Newsroom/DomesticApproachtoNationalIntelligence.PDF.

264 "How to Do It? A Different Way of Organizing the Government," 9-11 Commission Report, National Commission on Terrorist Attacks upon the United States, https://govinfo.library.unt.edu/911/report/911Report_Ch13.htm.

265 "In Domestic Intelligence Gathering, the FBI Is Definitely on the Case," *FBI*, Mar 21, 2007, https://archives.fbi.gov/archives/news/pressrel/press-releases/in-domestic-intelligence-gathering-the-fbi-is-definitely-on-the-case.

Chapter 21

266 M. R. Keogh-Brown, et al., "The macroeconomic impact of pandemic influenza: estimates from models of the United Kingdom, France, Belgium and The Netherlands,". *The European Journal of Health Economics* 11(6): 543-554, 2010.

267 V. J. Lee, et al., "Economics of neuraminidase inhibitor stockpiling for pandemic influenza, Singapore," *Emerging infectious diseases* 12(1): 95, 2006.

268 "Pandemic Emergency Financing Facility: Frequently Asked Questions," *The World Bank*, May 9, 2017, Accessed December 17, 2015, http://www.worldbank.org/en/topic/pandemics/brief/pandemic-emergency-facility-frequently-asked-questions.

269 "The economic impact of the 2014 Ebola epidemic: short and medium term estimates for West Africa," *World Bank Group:* Washington, DC., accessed November 16, 2015, http://documents.worldbank.org/curated/en/2014/10/20270083/economic-impact-2014-ebola-epidemic-short-medium-term-estimates-west-africa.

270 G. Verikios, et al., "The global economic effects of pandemic influenza. Report for 14th Annual Conference on Global Economic Analysis, Venice: Italy," accessed October 10, 2015 (Available from: http://static.rms.com/email/documents/liferisks/papers/the-global-economic-effects-of-pandemic-influenza.pdf.

271 A. Gale, "Fear of MERS poses risks to South Korea's economy," *The Wall Street Journal* June 10, 2015, accessed November 1, 2015, http://www.wsj.com/articles/fear-of-mers-risks-infecting-south-koreas-economy-1433928403.

272 S. M. Bartsch, K. Gorham, B. Y. Lee. "The cost of an Ebola case," *Pathogens and Global Health* 109(1): 4-9, 2015.

273 M. I. Meltzer, N. J. Cox, K. Fukuda. "The economic impact of pandemic influenza in the United States: priorities for intervention," *Emerging Infectious Diseases* 5: 659-671, 1999.

274 C. Becker, "Influenza economics. Providers and suppliers who usually reap big profits during flu season might find that a pandemic could backfire on their bottom lines," *Modern Healthcare* 35(45): 6-7, 2005.

275 "Pandemics Overview," *World Bank*, accessed October 10, 2015, http://www.worldbank.org/en/topic/pandemics/overview.

276 J. Otte, et al., "Impacts of avian influenza virus on animal production in developing countries," *CAB Reviews: Perspectives in Agriculture, Veterinary Science, Nutrition and Natural Resources* 3(080): 1-18, 2008.

277 P. B. Dixon, et al., "Effects on the US of an H1N1 epidemic: analysis with a quarterly CGE model," *Journal of Homeland Security and Emergency Management* 7(1): 1-16 2010.

[278] D. Langton, "Avian Flu Pandemic: potential impact of trade disruptions," The Library of Congress Congressional Research Service, 2006, accessed November 14, 2015, http://fpc.state.gov/documents/organization/68827.pdf.

[279] C. D. Chen, et al., "The positive and negative impacts of the SARS outbreak: a case of the Taiwan industries," *The Journal of Developing Areas* 43(1): 281-293, 2009.

[280] V. Marino, "SARS doubles demand for respiratory masks," *New York Times*, April 6, 2003, accessed November 20, 2015.http://www.nytimes.com/2003/04/06/business/yourmoney/06BDIG.html.

[281] S. Freisen, "The Impact of SARS on Healthcare Supply Chains," *Logistics Quarterly* 9(2): 10-11, 2003.

[282] "Ebola causes surge in sales of protective gear," *Chicago Tribune*, October 15, 2014, http://www.chicagotribune.com/news/local/breaking/ct-ebola-equipment-sales-met-20141015-story.html.

[283] P. M. Boffey, "C.D.C. Ebola guidelines aren't good enough for some," *New York Times*, October 30, 2014, http://takingnote.blogs.nytimes.com/2014/10/30/c-d-c-ebola-guidelines-arent-good-enough-for-some-states/?_r=0 .

[284] R. Daly, "Hospitals' Ebola preparation costs can vary widely," *Healthcare Financial Management* 68(12): 60-2, 2014.

[285] "Press Release: CDC increasing supply of Ebola- specific personal protective equipment for U.S. hospitals," Center for Disease Control and Prevention, Washington, DC, 2014, accessed October 20, 2015, http://www.cdc.gov/media/releases/2014/p1107-ebola-ppe.html.

[286] M. McCarter, "Emergency management/ disaster preparedness responders today: getting personal," *Homeland Security Today*, January 8, 2008, http://www.hstoday.us/focused-topics/emergency-managementdisaster-preparedness/single-article-page/responders-today-getting-personal/00989504d4ef44d78b6fee8e3a515a6c.html .

[287] D. Hinshaw, J. Bunge, "U.S. buys up Ebola gear, leaving little for Africa," *The Wall Street Journal*, November 24, 2014, http://www.wsj.com/articles/u-s-buys-up-ebola-gear-leaving-little-for-africa-1416875059.

[288] "#49 Medline Industries," *Forbes*, 2014, accessed November 20, 2015, http://www.forbes.com/companies/medline-industries/.

[289] "Corporate facts," *Medline*, 2014, accessed November 18, 2015, https://www.medline.com/about-us/key-facts.

[290] M. Krantz "Stocks involved with Ebola," *USA Today Money* October 14, 2014, http://americasmarkets.usatoday.com/2014/10/14/how-to-profit-from-ebola/?sf32477333=1.

[291] "LAKE Company Financials Income Statement," *Nasdaq*, accessed October 15, 2015, http://www.nasdaq.com/symbol/lake/financials?query=income-statement.

[292] "Alpha Pro Tech 2014 Annual Report," *Alpha Pro Tech*, Markham, Ontario, accessed November 1, 2015, http://www.alphaprotech.com/userfiles/doccenter/2014%20Annual%20Report%20with%20Bookmarks.pdf.

[293] R. Abrams, "Demand jumps for protective equipment as Ebola cases spur hospitals into action," *New York Times*, October 22, 2014,http://www.nytimes.com/2014/10/22/business/demand-jumps-for-protective-equipment-as-ebola-cases-spur-hospitals-into-action.html?_r=3.

[294] T. Crampton, "Nothing like deaths to sell life insurance: some manage to profit by SARS," *New York Times*, May 28, 2003. http://www.nytimes.com/2003/05/28/news/28iht-innovate.html.

295 A. Abelin, et al., "Lessons from pandemic influenza A(H1N1): The research-based vaccine industry's perspective," *Vaccine* 29: 1135-1138, 2011.

296 R. P. Velasco, et al., "Systematic review of economic evaluations of preparedness strategies and interventions against influenza pandemics," *PLoS One* 7(2): e30333-e30333, 2012.

297 "What would happen if we stopped vaccinations," Center for Disease Control and Prevention. 2014, accessed November 12, 2015, http://www.cdc.gov/vaccines/vac-gen/whatifstop.htm .

298 P. Flynn, "The handling of the H1N1 pandemic: more transparency needed," Council of Europe Parliamentary Assembly, Social Health and Family Affairs Committee: United Kingdom, AS/Soc (2010) 12, accessed November 22, 2015, http://assembly.coe.int/CommitteeDocs/2010/20100604_H1N1pandemic_e.pdf

299 D. Fahmy, "Drugmakers, doctors rake in billions battling H1N1 flu," *ABC News,* October14, 2009, http://abcnews.go.com/Business/big-business-swine flu/story?id=8820642

300 S. A. Plotkin, "Why certain vaccines have been delayed or not developed at all," *Health Affairs* 24(3): 631-634, 2005.

301 D. F. Maron, "Cross-border Ebola outbreak a first for deadly virus," *Scientific American,* July 30, 2014, http://www.scientificamerican.com/article/cross-border-ebola-outbreak-a first-for-deadly-virus.

302 A. Batson, "The problems and promise of vaccine markets in developing countries," *Health Affairs* 24(3): 690-693, 2005.

303 "Kalorama: Vaccines a 255 Billion Dollar Business in 2014," *PR Newswire,* February 13, 2015, http://www.prnewswire.com/news-releases/kalorama-vaccines-a-255-billion-dollar-business-in-2014-300035312.html.

304 R. D. Balicer, et al., "Cost-benefit of stockpiling drugs for influenza pandemic," *Emerging infectious diseases* 11(8): 1280-1282, 2005.

305 A. Patel, S. E. Gorman, "Stockpiling antiviral drugs for the next influenza pandemic," *Clinical Pharmacology & Therapeutics: Emerging Infections*, 86: 241-243, 2009.

306 D. G. McNeil, "Wary of attack with smallpox, U.S. buys up a costly drug," *The New York Times* March 13, 2013. (Available from: http://www.nytimes.com/2013/03/13/health/us-stockpiles-smallpox-drug-in-case- of-bioterror-attack.html?_r=0.

307 L. Mei, et al., "Changes in and shortcomings of control strategies, drug stockpiles, and vaccine development during outbreaks of avian influenza A H5N1, H1N1, and H7N9 among humans," *Bioscience Trends* 7(2): 64-76, 2013.

308 "Roche 2005: Record sales and operating profit. Basel: Switzerland," Hoffman La-Roche Ltd., 2006, accessed November 2, 2015, http://www.roche.com/media/store/releases/med-cor-2006-02-01.htm.

309 F. Godlee, "Conflicts of interest and pandemic flu," *British Medical Journal,* Jun 3:340: c2947, 2010.

310 J. E. Harrington, E. B. Hsu, "Stockpiling anti-viral drugs for a pandemic: The role of manufacturer reserve programs," *Journal of Health Economics* 29(3): 438-444, 2010.

311 P. Pongcharoensuk, et al., "Avian and pandemic human influenza policy in South-East Asia: the interface between economic and public health imperatives," *Health Policy and Planning* 27:374-383, 2011.

312 "Tekmira Announces Launch of Arbutus Biopharma, a Hepatitis B Solutions Company," *Arbutus Biopharma,* 2015, accessed October 5, 2015, http://investor.arbutusbio.com/releasedetail.cfm?releaseid=922758.

313 J. Shmuel, "How Ebola could affect your stock portfolio," *Financial Post*, October 2, 2014, http://business.financialpost.com/investing/how-ebola-could-affect-your-stock-portfolio.

314 A. Pollack, "Sales of U.S. company's drug rise as Chinese try It against SARS,". *The New York Times*, May 6, 2003. http://www.nytimes.com/2003/05/06/world/sars-epidemic-treatment-sales-us-company-s-drug-rise-chinese-try-it-against-sars.html .

315 A. Ohlheiser, "When Ebola comes to the U.S., who stands to profit?" *The Washington Post* October 1, 2014, https://www.washingtonpost.com/news/to-your-health/wp/2014/10/01/when-ebola-comes-to-the-u-s-who-stands-to-profit/.

316 D. Martin "How coca traders make money on the Ebola scare," *PBS Newshour: Making Sense* October 28, 2014, http://www.pbs.org/newshour/making-sense/cocoa-traders-make-money-ebola-scare/.

317 S. Begley, "Flu-conomics: The next pandemic could trigger global recession," *Reuters* January 21, 2013, http://www.reuters.com/article/2013/01/21/us-reutersmagazine-davos-flu-economy idUSBRE90K0F820130121#Hv9LzZz6o7z8mrmT.97.

318 S. Cooper, D. Coxe, "An Investor's Guide to Avian Flu," BMO Nesbitt Burns Research: Toronto, 2005.

319 S. Broyer, C. Brunner, "Pandemic: A short guide for investors," *Natixis Flash Economics* (343):1-9, 2009, accessed November 22, 2015. http://cib.natixis.com/flushdoc.aspx?id=48104.

320 Visiongain, About Us, accessed October 28, 2015, https://www.visiongain.com/Content/3/About-Us.

321 "Merchandise trade (% of GDP)," World Bank Data Bank, accessed November 15, 2015, http://data.worldbank.org/indicator/TG.VAL.TOTL.GD.ZS?page=1.

322 P. N. Fonkwo, "Pricing infectious disease," *European Molecular Biology Organization Reports: Science and Society* 9(1S): S13-S17, 2008.

323 Food and Agriculture Organization of the United Nations, Animal Production and Health Commission for Asia and the Pacific. 2002. Manual on the diagnosis of Nipah virus infection in animals. Food and Agriculture Organization Regional Office for Asia and the Pacific. Publication no. 2002/01, accessed December 1, 2015, ftp://ftp.fao.org/docrep/fao/005/ac449e/ac449e00.pdf.

324 United States Census Bureau, Foreign Trade: Top Trading Partners- September 2015, accessed November 1, 2015, https://www.census.gov/foreigntrade/statistics/highlights/toppartners.html#imports.

325 K. E. Jones, et al, "Global trends in emerging infectious diseases," *Nature* 451(7181): 990-993, 2008.

326 "Business Alert—China: Impact of SARS on Chinese Economy," Issue 06, 2003, *Hong Kong Trade Development Council*, accessed October 1, 2015, http://info.hktdc.com/alert/cba-e0306sp-4.htm.

327 L. Herbert, "Trading on quality and disease-free in the Middle East beef market," *ABC Rural* April 29, 2014, http://www.abc.net.au/news/2014-04-29/marketing-aussie-beef-to-middle-east/5417236.

328 "FTC and FDA crack down on internet marketers of bogus SARS prevention products," Federal Trade Commission, 2003, accessed December 1, 2015,https://www.ftc.gov/news-events/press-releases/2003/05/ftc-and-fda-crack-down-internet-marketers-bogus-sars-prevention.

329 "WGA Launches Ebola Pandemic Response Product for Loss of Income," William Gallagher Associates: Boston, MA, 2014, accessed November 10, 2015, http://www. wgains.com/launches-pandemic-response.

330 A. Whiting, "New pandemic insurance to prevent crises through early payouts," *Reuters*, March 26, 2015, http://www.reuters.com/article/2015/03/26/us-global-pandemics-ins uranceidUSKBN0MM1XD20150326#E7CZOsAZOHPBPgki.97.

Chapter 22

331 World Health Organization. International Health Regulations (2005). Geneva: WHO; 2008.

332 A. H. Maslow, "A theory of human motivation," *Psychol Rev*. 1943 Jul;50(4):370.

333 A. Prüss-Üstün, R., et al., "Safer water, better health: costs, benefits and sustainability of interventions to protect and promote health," *Geneva: World Health Organization*; 2008.

334 "Health through safe drinking water and basic sanitation," World Health Organization, Geneva: WHO; 2015 [cited 2016 February 9]. http://www.who.int/water_sanitation_health/mdg1/en/.

335 H. Cooley, et al., "Global Water Governance in the 21st Century," Oakland: Pacific Institute; 2013 Jul.

336 R. G. Garriga, A. J. de Palencia, A. P. Foguet, "Improved monitoring framework for local planning in the water, sanitation and hygiene sector: from data to decision-making," *Sci. Total Environ*. 2015 Sep 1;526:204-14.

337 V. Curtis, S. Cairncross, "Effect of washing hands with soap on diarrhea risk in the community: a systematic review," *Lancet Infect Dis*. 2003 May 31;3(5):275-81.

338 S. P. Luby, et al., "Effect of handwashing on child health: a randomised controlled trial," *Lancet*. 2005 Jul 22;366(9481):225-33.

339 R. Lenton, A. M. Wright, K. Lewis., "Health, dignity, and development: what will it take?," London: United Nations Development Program, UN Millennium Task Force on Water and Sanitation 2005; 2005.

340 "Children: reducing mortality," Geneva: WHO; 2016, accessed January 2016, http://www.who.int/mediacentre/factsheets/fs178/en/.

341 D. L. Goodman, H. van Norton, "Water, sanitation and hygiene education for schools: roundtable proceedings and framework for action," Oxford: United Nations Children's Fund and IRC. 2005.

342 G. Lofrano, J. Brown, "Wastewater management through the ages: a history of mankind,". *Sci Total Environ*., 2010 Oct 15;408(22):5254-64.

343 K. Watkins, "Human Development Report 2006 - Beyond scarcity: power, poverty and the global water crisis," *UNDP Human Development Reports* (2006), Nov 9, 2006.

344 C. Tacoli, "Urbanization, gender and urban poverty: paid work and unpaid carework in the city," Human Settlements Group, International Institute for Environment and Development; Mar 2012.

345 L. Rodríguez, E. Cervantes, R. Ortiz, "Malnutrition and gastrointestinal and respiratory infections in children: a public health problem," *Int J Environ Res Public Health*, 2011 Apr 18;8(4):1174-205.

346 K. Sorsdahl, et al., "Household food insufficiency and mental health in South Africa," *J Epidemiol Community Health*. 2011 May 1;65(5):426-31.